Communications for ICT

The Essential Guide

David Tuffley

To my beloved Nation of Four
Concordia Domi – Foris Pax

Your audience is one single reader. I have found that sometimes it helps to pick out one person -- a real person you know, or an imagined person and write to that one.

-- John Steinbeck

Acknowledgements

I gratefully acknowledge the *Turrbal*, *Jagera* and *Noonuccal* indigenous peoples, on whose ancestral land I write this book..

Contents Summary

High-level summary of contents (detailed Contents follow)

3

Detailed Contents

Detailed summary of contents:

11

Preface

Communications for ICT is an essential guide to technical communication for IT professionals. This useful text goes well beyond the conventional technical communication text book by also including important topics such as professional conduct, argument/persuasion, software project documentation (like requirements specifications), international communication, high-performance project teams and the ethical technologist. These are in addition to how to write technical reports, user manuals and whitepapers.

Communications for ICT supports a course in the School of ICT at Griffith University, a top-ten research university in Australia. The author combines 15 years of practical experience as a technical writer in the IT industry, with his ongoing academic teaching and research since 1999. Dr. Tuffley has produced a practical guide that can be understood and applied by any IT industry worker wishing to develop these important, but often neglected skills.

Offering outstanding value for money, Communications for ICT provides you with instruction that is the equivalent of a high-quality university level, education. It is the prescribed text book for a university course by the same name.

Main topics:

- Principles of clear communication
- Writing university assignments
- Technical communication process
- Writing software user documentation, technical reports
- Modes of Technical Communication (paper-based, on-line)
- Tool support
- Editing techniques
- Cross-cultural communication
- Presenting a persuasive, well-argued case
- Effective team membership
- The ethical technologist

About the Author: David Tuffley PhD is a Lecturer in the School of ICT at Griffith University in Australia. He is also a Senior Consultant with the Systems & Software Quality Institute, a systems engineering best practice organisation based in Brisbane, Australia. David worked for 20 years in the IT industry as a technical writer, business analyst and software quality consultant. David has a successful track record of writing and publishing IT-related books since 1993.

Introduction

Effective communication in the workplace and in life generally is a major factor contributing to career success and happiness in life. This course introduces you to the principles of clear communication across the areas that a technology-focussed worker will need.

The book takes a holistic approach by placing the content in its larger context of a person's life. You are shown how these principles apply to your work-related activities, but also how they apply in how you think and live.

The book therefore aims to equip you to become a competent technical communicator with a good understanding of the techniques and tools that are used to produce a broad spectrum of reader-friendly documentation. The book also includes a component on the history of technology, effective team communication skills and the basics of ethical IT

Technology in Perspective

For hundreds of thousands of years, humans have excelled as tool-makers. One of our earliest technologies was the ability to communicate, which is the development of language and writing. With these our ancestors were able to coordinate activities such as hunting and gathering that allowed them to survive in a generally hostile world.

Then came the many hand-tools and weapons that further extended our capabilities to obtain food and defend ourselves. In the present day, information technology can be seen as the most sophisticated tools we have ever made. More than just inanimate objects though, information technology is an extension of our minds. Technology lets us extend our ability to think and process information beyond our biological brain, out into the environment. Just think of how you use your computer to process information for you.

This ability to extend our minds into our tools did not begin with information technology. We have always done this. Andy Clark, a respected cognitive scientist reports that brain scans show that if you were to pick up, for example, a garden rake and start to use it to gather leaves, within a short time your brain would have mapped the tines of the rake to be extensions of your hands. We call it haptic touch.

The computers that we have come to depend on are just another tool that we project our minds into and use to outsource some of our thinking tasks. If you doubt this, imagine if you lost your personal computer. In some ways, it would be like having a stroke. Part of your brain would have disappeared, and you would very much feel the lack of it. You might feel lost and debilitated until a replacement is found, complete with restored data.

So people have a closer relationship to information technology than is commonly realised, having become an extension of our biological mind. As millions of extended minds have reached out and merged with each other we can observe a remarkable phenomenon, the formation of a new layer of consciousness in the world.

Language is the currency of thought

George Orwell's classic novel *1984* warned us that a person can only think what they have language to think with. Language is the currency of thought. A person with a limited vocabulary and an inability to construct meaningful language therefore has very limited capacity to think.

In *1984*, the government reduced the number of words people were allowed to use, and in so doing limited their ability to think critically. This suited the government in the novel, who wanted to exercise complete control over people's thoughts and actions.

This course aims to do the opposite; to give you the tools with which to think positively and clearly. It is only from such a mind-set that clear and effective communication can come. It will deal with the psychology of language as well as the techniques of language. In approaching it this way, it matters less whether English is your first language.

Code of Professional Conduct

This section is a detailed introduction to the topic of how IT professionals must behave in the work environment in order to be perceived as professionals. This topic will be advanced through out the remaining chapters.

For IT practitioners, behaving in an ethical, professional manner involves more than simply learning a code of conduct. To be ethical, a practitioner must be competent and highly effective. These are the foundation of ethical IT practice. How can an incompetent person behave ethically? They are unable to perform the work.

Ethical IT practice therefore involves being competent, and also to develop a success mentality that works proactively to be highly effective.

Every developed country has a code of professional conduct for IT professionals. There are also codes that apply internationally, put forward by professional bodies like IEEE that have a global reach.

The Australian Computer Society's Code is arguably as good a code as any to and a good place to begin our exploration of ethical IT practice..

The following summary is provided with thanks to ACS (full version: http://www.acs.org.au/index.cfm?action=show&conID=copc).

The Public Interest

Safeguard the interests of your clients provided that they do not conflict with the duties and loyalties owed to the community, its laws and social and political institutions

In performing work for a client your priority should be to satisfy that client's needs and to meet the specifications to which you are committed. If, however, in meeting these requirements you are forced to breach law or inflict damage upon a third party, then you are professionally responsible to make the client aware of these consequences and agree an alternative course of action.

Integrity

Do not breach public trust in the profession or the specific trust of your clients and employers.

Observance of utmost honesty and integrity must underlie all your professional decisions and actions. Circumstances will undoubtedly arise during the course of your professional career where it may appear to be beneficial for you to deceive your client in some way. The resultant short term gains from this type of behaviour is not acceptable professional practice, nor is it worth eroding the confidence and trust that is built up over the longer term.

Confidentiality

You must not disclose information acquired in the course of your professional work except where consent has been obtained from the rightful legal owner or where there is a legal or professional duty to disclose

This is applicable to most professions, but it is particularly applicable to you as an Information Technology professional as you are likely to have access to clients' information due to the nature of your work. You should be aware that information is the property of the client, and must not be distributed freely or used for your personal advantage or that of a third party without the client's consent.

Objectivity and Independence

Be objective, impartial and free of conflicts of interest in the performance of your professional duties

In each professional assignment undertaken, you must be seen to be free of any interest which is incompatible with objectivity. Always make sure you are aware of your client's objectives and the benefits he is looking for, and be careful not to lose objectivity created by the latest development technology or by the desire to promote your own product.

In the situation where a conflict exists between two or more clients, a full and frank explanation and disclosure of the conflict should be made to the clients.

Competence

Accept only such work as you believe you are competent to perform and do not hesitate to obtain additional expertise from appropriately qualified individuals where advisable

You should always be aware of your own limitations and not knowingly imply that you have competence you do not possess. This, of course, is distinct from accepting a task of which the successful completion requires expertise additional to your own. You cannot possibly be knowledgeable on all facets of Information Technology but you should be able to recognise when you need additional expertise and information.

Keeping Up-To-Date

Keep yourself, and subordinates, informed of such new technologies, practices and standards as are relevant to your duties Others will expect you to provide special skills and advice; and in order to do so, you must keep your knowledge up-to-date. This is true for members of all professions, but particularly so in Information Technology which is developing and changing rapidly.

You must also encourage your staff and colleagues to do the same, for it is impossible to retain one's professional standing by relying only on the state of one's knowledge and competence at the time professional status is achieved.

Subordinates

Ensure subordinates are trained in order to be effective in their duties and to qualify for increased responsibilities

Take action to ensure that your hard won knowledge and experience are passed on in such a way that those who receive it not only improve their own effectiveness in their present positions but also become keen to advance their careers and take on additional responsibilities.

Responsibility To Your Client

Actively seek opportunities for increasing efficiency and effectiveness to the benefit of the user

Whatever the precise terms of your brief, you should always be aware of the environment surrounding it and not work solely towards completion of the defined task. You must regard it as part of your duty to make your client aware of other needs that emerge, unsatisfactory procedures that need modification and benefits that might be achieved. You, as an innovator, should take into account the relevance of new methods and should always be looking for the possibility of additional benefits not foreseen when the project was planned.

You should also look beyond the immediate requirements to the needs of the ultimate user. For example, the invoice your system produces may be right for company accounting procedures but confusing for the person who is being asked to pay against it.

Promoting Information Technology

Endeavour to extend public knowledge, understanding and appreciation of Information Technology People, for various reasons, can often be mistrustful or demonstrate resistance when it comes to Information Technology. Aim to promote Information Technology by educating people as to the benefits that can be achieved through its application to their business. You should, however, only express an opinion on a subject within your level of competence and when it is founded on adequate knowledge and honest conviction, and oppose any untrue, inaccurate, exaggerated or misleading statement or claims.

The Image Of The Profession And The Society

Refrain from any conduct or action in your professional role which may tarnish the

image of the Information Technology profession or unjustifiably detract from the good name of your professional body Information Technology is a relatively new industry, characterised by rapid change. It is unlike other professions such as accounting or medicine in that it has not had the opportunity to evolve over many years and acquire its own standards and legislation.

Any person can, in fact, practice within the industry and claim the title of Information Technology Professional.

The Society is aware that the public image of Information Technology is of some concern. It is imperative, therefore, that members of the Information Technology profession endeavour to maintain a professional standard that improves and enhances the industry's image.

Writing white papers

A white paper (or "whitepaper") is a well-researched, expository piece of writing that authoritatively reports on issues of interest or concern. White papers are used to inform interested parties, and often to help them make decisions.

This chapter outlines a proven method for writing high quality white papers for the professional world. The approach can be equally used when writing university

Many books have been written on this topic. Some go into much detail. This chapter presents the essence of report writing, a stripped down, easy to understand and apply method for writing high quality reports, even if you do not consider yourself a good writer.

Time Management

Time is a valuable resource. Once spent or wasted, it can never be recovered. All of us have a limited amount of time on this earth, even if it seems a long time when you are young.

People who learn how to manage their time to greatest effect are among the most successful people in the world. It is an essential skill for success. If you spend your time wisely, putting your efforts into those activities that produce the greatest return is a success secret that many people never fully appreciate. On the negative side, poor time management is a sure way to consign you to a mediocre life of low-achievement.

Many people arrive in the IT world without having mastered this essential skill while at university or college. Traditional IT education focuses heavily on the technical side of knowing how to write software but not reports. It is essential that you learn this skill now. It will make your life a whole lot easier.

Time management can be simplified to a few essential points.

Use a Scheduling Tool

Use a scheduling tool (like Google Calendar, or similar). At the beginning of the project, enter all tasks into the calendar. You must be able to see at a glance, everything that needs to be done, weekly and monthly.

Put other important dates into your schedule, like when you will be travelling, social engagements, sporting events, holidays and recreation. Everything.

Look at your calendar every day to make sure you know exactly what is going on, and what needs to be done in the days and weeks ahead.

Expect the Unexpected

Give yourself an extra week or two to complete assignments by starting them early. Leaving an assignment until a few days before it is due, then finding that you are unexpectedly busy, sick or otherwise occupied will put you between a rock and a hard place.

Most managers will not grant an extension to the submission deadline on these grounds, as it is simply poor time management that created the problem. You have known about the task for weeks. After all, you are a professional and are expected to behave like one.

In the professional world, missing deadlines for any reason is considered unprofessional. Disorganised people quickly earn a reputation for being unreliable, and this severely limits their career prospects. Indeed, in any profession, and beginning with your time at University, you are expected to have reached a level of control over your own life that allows you to reliably deliver on time what it is you must deliver.

Break tasks down into chunks

Assignments can be intimidating. The size and scope of some assignments seems so large that it can seem very daunting.

The best approach for any complex task is to break it down into smaller pieces. Then deal with each piece without worrying about the remaining pieces until it is time to do them. If you give yourself enough time, and work steadily away at it until the job is finished, you can produce almost anything.

For example, after having been a first year student, a PhD student has worked their way up to writing a 100,000 word thesis describing a piece of original research that is fit for publication in the world. This might take him/her four years to complete, but planning the whole project in detail, breaking it down and completing each task in turn, the job gets done.

This is how anyone completes any major work. The hard part is cultivating enough self-discipline to keep working when there are so many more pleasant things you can think of to do. Enter the world of procrastination, a student's biggest enemy.

Procrastination is your enemy

Procrastination is an all-too-human weakness that causes people to spend time on less-important activities, at the expense of the more important activities.

To combat procrastination, you objectively look at what needs to be done, then decide how best to spend your time. Try to identify those activities that bring the greatest benefit and do them. The less important activities can either be done later, or not at all.

Guide to effective report writing

When you wish to instruct, be brief; that men's minds take in quickly what you say, learn its lesson, and retain it faithfully. Every word that is unnecessary only pours over the side of a brimming mind.
-- Cicero Roman author, orator, & politician (106 BC - 43 BC)

This section outlines a practical method for students and professionals in a wide range of disciplines to develop reports (including essays) that address the needs of the reader and which are expressed in language easily understood by the reader.

The skills you learn here will be directly transferable to the professional world.

Note that a "Report" is a generic term covering any piece of expository writing, including the essays you write at University. We are teaching you how to write reports for work by practicing on University essays.

The design of effective reports

Designing effective reports is a three stage process:

Planning

Planning is the most critical stage for the success of the document. Planning means thinking about the following:

1. **Who will be your audience?** Describe the responsibilities, job level(s) and attitudes of all primary readers. Who will act on what you write? Who else might read what you write? What was the situation that led to the need for this document?

2. **What is your purpose in writing the document?** Why are you writing? What do you want to achieve with your writing? Do you need to restate your purpose to

fit the reader's perspective? What kind of image do you want to project of yourself.

3. **How will you organize your message for greatest effectiveness?** What main ideas do you want to include? Which ones to exclude? Considering your reader(s) attitudes, needs and perspectives, in what order do you think these ideas should be presented?

4. **Visual aids**. What visual aids can you use to make the message as concise and clear as possible?

5. **Writing style**. What writing style will be most effective? Based on your analysis of who will read your document, what language are they most likely to understand.

Writing, evaluating and revising

This applies the results of your analysis in stage 1 to the development of your document (which could be a report, proposal, procedure, policy or even an article for publication). As you write, you need to constantly evaluate what you have written against the decisions made in stage 1. You will revise what you write. For important documents, you need to set time aside to evaluate and revise what you have written.

Editing

Editing focuses on the mechanics of your writing to ensure technical accuracy and general clarity.

While a three stage process may seem to be a burden at first, with practice you'll find that it becomes an efficient, reliable guide to developing any document - telling you what you need to say and not say.

Understanding the business communication context

For many people, their formal writing style is derived, or is at least influenced, by the kind of writing they did at school and university. There are a number of important distinctions that need to be made between university writing and writing in a professional context.

In understanding the context of business writing and the distinction between academic and business writing, the following points need to be addressed:

Writing at work achieves job goals

Your work writing helps the organization achieve its corporate goals and allows you to do your job - its a prerequisite for job effectiveness. It is not a vehicle to convince anyone that you know a subject and deserve at least a passing grade.

Writing at work addresses a variety of readers

The reader for whom you write is no longer a single reader (a stakeholder, a manager, a professor etc.). Surveys have found that business writings are addressed to many readers with varying educational backgrounds, but most often to people who know less about the subject area than the writer themself. You will often need to write for people in other parts of the organization. They will read what you write based on their own jobs, backgrounds, education and technical expertise. Unlike the people you were at school or university with, those with whom you work will probably not have the same educational and technical background as you. Your success at communicating with them will depend largely on how well you gauge the unique background of your reader(s).

Writing at work addresses readers with different perspectives

Your reader, unlike your teacher, will feel no obligation to read what you have written. They have their own jobs and will seldom bother to read something unless they see it as being helpful to them personally.

Writing at work creates excessive paperwork

We live in the age of infoglut where the quantity of information available to us increases exponentially. People have more to read than they'll ever have time to read. Few documents are read completely - most are skimmed. So when a reader picks up your document, they will be asking themselves; What is this? Why should I read it? How does it affect me? What am I going to have to do? They will want to find the main points straight away. They will become impatient if they can't find them by glancing down the pages. In short, your reader is not a captive audience - you have to make your message clear and easy to read. You also need to make your message as interesting an relevant as possible. Correct writing that cannot be read quickly and easily is not

likely to be read. Few readers will be impressed with verbose, disorganized writing, even if each sentence is technically correct.

Writing at work may be read by readers unknown to the writer

Your writing will be read by two categories of people: the primary readers for whom it was intended, and secondary readers who will perhaps receive copies because the information is relevant to their jobs. Whoever your primary readers are, you should always anticipate the secondary readers. Its important not to underestimate the problem presented by unknown readers. Copies of your reports will often be placed in files that can be accessed by people who know little or nothing about the situation you are writing about. Yet these same reports can be used in assessing your performance and determining your promotion potential. Its often the case that readers who know little about you will make decisions about your competence based on how well you write. In other words, what you write at work is much more than a knowledge indicator that helps a teacher to determine a grade.

Writing at work has an indefinite lifespan

Much of what is written in organizations will remain in the organization's files for years. Because known and unknown readers may use your documents for an indefinite period, the problem of trying to identify the reader is further complicated by the length of time that the document is accessible. Academic writing, on the other hand, applies to a specific semester and lecturer.

Writing at work can create legal liability for the writer and the reader

Unlike academic writing, professional writing has the potential for being used against you in court. Both the writer and the person approving the document become responsible for the content. You need to keep in mind the indefinite lifespan of the material and the difficulty of knowing just who will read the document. In addition to this, you need to also remember that people may use your writing for reasons you never considered. Its possible that people might take sentences and even paragraphs out of context for use in situations unrelated to your original intent. They can then use what you say to support claims against you and your organization. Because we live in an increasingly litigious society, its important to consider these legal aspects.

How to organize business communication

Organization can help or hinder your reader in understanding your writing. To improve the accessibility of your writing, do the following:

1. Use clearly worded titles.
2. Use a summarising opening.
3. Use headings to indicate partitioning of ideas.

4. Use topic sentences to announce the content of paragraphs to help the reader move quickly through.

By following the above guidelines, you enable the reader to quickly answer the four basic questions: What is this? Why am I receiving it? How does this document affect me? What am I supposed to do? The guidelines allow a busy reader to navigate easily and grasp the essential content without needing to read the whole document.

The following specific guidelines can be used to achieve the above:

Structure the content to help the reader find key points

As indicated in the introduction above, when the average reader picks up a document, they have four basic questions in mind: What is this? Why am I receiving it? How does this document affect me? What am I supposed to do?

The structure and internal organization must help readers find the answers to these questions quickly. The ease with which the reader can grasp the essential meaning of the document determines how much of the document they decide to read.

Begin with the most important information

Once you know what you want your document to accomplish and what your readers need from the document, place that information at the beginning in a paragraph for medium length reports and in an introduction or summary for long report or reports that will be read by a number of readers who have different levels of familiarity with the subject.

Discussion, development or support of the main information follows in the body of the document. The conclusion follows the body. The following structure is appropriate when the reader is mainly interested in the conclusion, recommendation or result being reported upon.

1. **Main ideas** - paragraph summary or introduction plus paragraph summary.

2. **Development** - explanation, support for each idea given in the summary.

3. **Conclusion** - restatement of main ideas, what the reader needs to do in response to the main information.

Attachments - supporting data

When your report focuses on analysis as the heart of the report, or you need to emphasize the discussion in order to justify the conclusion, or if the reader is likely to be hostile to your conclusion and you need to change the reader's attitude to the subject before presenting the conclusion, then the following structure is appropriate.

1. **Main ideas** - abstracts, summary, foreword.

2. **Development** - discussion, main body of report.

3. **Evaluation, Conclusions, Recommendations** - results of the discussion.

4. **Appendix**, **Attachments** - supporting data for ideas in the discussion.

Design clear titles

Formal reports should use titles that are both concise and descriptive of content. Research backs up what common sense suggests, that readers comprehend documents more quickly if they know what the document is going to discuss.

Design the summary

While the summary is placed first in a document, it is usually the last item you write. You are best able to encapsulate a topic after you have become thoroughly familiar with it during the development of the discussion.

A summary may consist of one sentence, one paragraph or several paragraphs for long reports. The summary should address the four questions alluded to previously: What is this? Why am I receiving it? How does this document affect me? What am I supposed to do? How long the summary is depends on the size and complexity of the report.

Design the introduction

Whereas summaries give the reader the main ideas presented in the document, introductions provide the setting in which writer prepares the reader for what is to follow in the discussion. Whether or not you provide an introduction depends on the readers familiarity with the subject. If the reader is familiar with the problem or situation that led to the document, you can open with a summary and omit the introduction.

If however you anticipate that some readers will not readily understand what the document is about and why it was written, an introduction will give them the background necessary to answer the four questions (see previous section). The introduction enables any reader, even one months or years in the future to understand why it was written. So if in doubt, include an introduction.

Like summaries, introductions vary in length and development, depending on the length and complexity of the document. As a general guide, the introduction should contain the following elements:

1. Subject of the document.

2. Purpose of the document.

3. Background - what led to the document.

4. Major points discussed.

5. Scope of document - topics covered and not covered.

6. Procedure for developing the content.

Research shows that readers understand content more quickly when they are prepared for that information, that is, if they are told up front what they are going to read. Introductions therefore improve the accessibility and readability of documents because they tell the reader what to expect.

Consider using overviews

Like introductions, overviews state the topic, the purpose of the information that follows, and how it is organized. Overviews frequently precede major sections, particularly major parts of the discussion of the report.

Like the introduction, the overview improves the readability of that section. Overviews are generally shorter than introductions because they deal only with the section that follows and helps the reader know how to approach that content. Overviews are particularly useful in long reports made up of several long sections.

Design the discussion

The main discussion section represents the bulk of the document. The discussion gives full development, explanation and support for the main ideas presented. But studies show that the discussion is the least read part of the report. This is basically because readers are busy and are only interested in the part(s) of the report that affects them, and read the parts that give them the information they need. For that reason, the discussion must be structured to enable the reader to find what they need without having to search through pages of information. In deciding how much to say in the discussion, keep your reader's knowledge level in mind.

When designing the discussion, be sure that the main points given in the summary are easy to locate in the discussion itself. Place the most important concepts first, unless your ideas follow a chronological or sequential order.

In developing the discussion, you will need to use the same development methods you have used previously at university - division and classification, description, illustration, comparison/contrast and causal analysis.

Design the conclusion

The final section, the conclusion, may include an evaluation, summary, and/or any recommendations based on ideas you have stated in your discussion. In the conclusion, you may reiterate your main points, elaborate any point that seems particularly important, and (if necessary) state any recommendations you believe are necessary.

Design attachments

Many times, you will include figures, data, calculations, or other information that support assertions made in the discussion. Including this supporting detail in the main part of the report is not appropriate since excessive detail hinders the reader's ability to follow your ideas.

If you do provide supporting material, you need to draw the reader's attention to it and indicate when it is relevant.

Structure paragraphs for easy reading

Readability research indicates that paragraphs are easier to understand when they begin with a topic sentence, rather than place the topic sentence in the middle or leave it out altogether.

The same principle applies to paragraph development as to document development generally - place the most important information at the beginning. When developing paragraphs, begin with a topic sentence that summarises the idea that will be discussed in that paragraph. Just as the report summary and introduction prepare the reader how to read the report, so the topic sentence prepares the reader for the information that follows.

Presenting information visually

Presenting information visually comprises the following:

The value of graphics

Considering using graphics increases the time it takes to produce a document, you might be tempted to ask: "why do I need to bother?" Graphics can increase the efficiency with which your reader extracts meaning from your document, intensifies the impact of the message, and improves the retention of information.

Ours is a visual society. We live in an age of too much information, an age when people have more to read than they can possible manage, where work has become increasingly abstract and data-defined, where pictures are as important as words, and where documents compete with other documents for the reader's attention. Graphics help the reader access meaning quickly and improves the chances that your document will be read and acted upon.

Thinking visually

Software packages like Excel, Powerpoint, and Photoshop offer a great deal of flexibility with the production of graphics. Now that producing graphics has been made easy, the most frequent problem is selecting a graphic that fits the information and meets the needs of the reader. For example, when showing comparative quantities, pie charts or bar graphs can be effective. But if a great deal of segmentation is required, a table may be the only choice (i.e. if there are 15 or more categories to display). Tables are however less effective than graphs at showing similarities and differences among quantities. One option is to use a combination of table and bar graph. Used together, the table gives the figures credibility, while the graph helps the reader see the relationships among the figures and gives them meaning. Where space is limited, the graph should be omitted in favour of the table.

Guidelines for writing visually

In short, writing visually helps you to be clear and convincing. It also helps to interest the reader in what you write by lessening the deadly effects of unbroken paragraphs.

The following guidelines will help you to decide when to use graphics:

1. If your writing includes extensive data.

2. If the document requires your reader's immediate attention, you can provide encouragement by helping

the reader see your ideas rather than having to read them.

3. If you consider the reader will not be motivated to read, visual aids will encourage them because seeing is easier than reading.

4. If the reader is not familiar with the topic, visual aids can enhance understanding, stimulate interest, and even improve retention of information.

5. The distracting the reader's environment, the more a visual display will help the reader to process your message.

6. Use tables if the reader understands the material; flowcharts if the material is complex; and short sentences (listed) if you want the reader to memorize and retain the list.

Effective style

An analysis of your readership, the purpose of the document, and the context in which it will be read will help you decide the content, organization, format and very importantly the best writing style to use. In making each decision, your purpose is to enable the reader to grasp the content quickly and easily, to elicit the response you want and to ensure that the reader has a positive attitude towards you.

Style includes the words and phrases you choose, the sentence structure you use and the way you express ideas.

Style can obviously vary widely and a good writer adjusts the style to the reader.

In determining the best style, consider the following:

1. Determine the reader(s) knowledge of the subject.

2. Determine the reader(s) expectations about style based on the kind of writing you are doing.

3. Determine the reader(s) probable reading level, based on the context in which the document will be read.

4. Determine your relationship to the reader and your identity as a writer.

Summary:

1. Know your reader's background and level of expertise on the topic. Use words commensurate with that level.

2. Anticipate the context in which the reader will read the document. Find out if a particular style is expected.

3. For routine writing, use a conversational style: write as you speak.

4. Remember the reader is human and will respond emotionally as well as intellectually to what you write. Be sure that your writing conveys a positive attitude to the material and respect for the reader.

5. Write as simply as possible. Most readers want information that is easy to read and understand. They are not impressed with pompous language that obscures, rather than clarifies meaning.

Determine the reader(s) knowledge of the subject

The reader's familiarity with the subject will determine how many specialized terms you can safely use without baffling the reader. If they are thoroughly familiar with he subject, you can use acronyms, jargon and other specialized terms. If the reader is not thoroughly familiar with the subject, you should limit the use of specialized words or phrases to those you are sure the reader will understand. If you are in doubt about your reader's knowledge on a subject, eliminate specialist language, define it, or substitute words or phrases that will clearly express your meaning.

Determine if a particular style will be expected

Many times, the person for whom you write - your manager, for example - will want you to write in a certain way, that is, to use specific phrases. In many organizations, documentation standards dictate the format and to some extent the content of documents (i.e. what information must be included on the front cover). Beyond adhering to the prescribed documentation standards, there is still scope to adapt your writing style to the reader. For example, the kind of style used for a report which will be read primarily by a high-level decision-making committee will be less technical than that used in a report primarily for consumption by technical people.

Determine reader(s) probable reading level

Reading comprehension also determines your choice of writing style. Reading comprehension level refers to the degree of difficulty in written material that a reader can accommodate without misunderstanding the content.

Determine your relationship to the reader

In analysing the reader, you need to determine the kind of image that you must project based on your position in the organization and your relationship to the reader.

In selecting the appropriate image to project, the language you choose can sound formal, informal, neutral, effusive, rude, dictatorial, or respectful. Your style can project a whole range on emotions upon the reader. So, when choosing a style, you need to be sure that it is appropriate to your position in the organization, to the content of the report, and to the readers.

Overcoming writer's block

We have all experienced it as the painful condition of having to complete a complex writing task, but being unable to get started and/or finish the job. Not to worry, help is at hand. With the right approach, almost any writing task can be managed.

Overcoming Writer's Block is a practical guide for anyone wishing to learn some practical strategies to deal with this all-too-common problem.

Preparation

The problem is often that you are expecting to hear the finished product being dictated in your mind by that mysterious process called inspiration. But before the words will start to flow you need to know enough about the subject.

There is an old saying, and true, that inspiration must be accompanied by much perspiration (perhaps not literally), but the point is clear that you have to put in quite a lot of work before the muse of inspiration will visit you and whisper those golden words in your ear.

So if you are experiencing writer's block, it's generally a sign that you don't yet know enough about the subject. There is no getting away from needing to spend more time getting to know the subject well.

Manage expectations

Part of the preparation process is to manage your expectations about how much time and effort is needed at this stage of writing. Through wishful thinking many people expect to complete this stage quickly and with minimal effort. But it simply does not work that way.

You need to accept that before you can start writing, you must first program your thinking with enough input material that you then have something to write about. And that of course takes time, probably more time than you want to spend. After all, there are so many other things you would like to be doing with your life besides writing this damn essay/report.

There is also the tendency to want to plunge into the writing task as soon as possible, to get the job done so you can then do other things that you enjoy.

So the key to good preparation is to have realistic expectations about how long the preparation stage will take. If you accept that this stage requires a fair amount of effort, you will be much more likely to do it well.

Incidentally, there is a big pay-off for good preparation. Studies have shown the "Return on Investment" (or ROI) on good preparation is something like 4:1, which means, for every hour you spend doing it, you save yourself four hours later in the project through not needing to re-do sections and fix mistakes. The fact is, and professional writers know this, that good preparation actually saves time and produces good results.

Structure

Most pieces of writing have a standard structure or organisation. It is the skeleton which you then flesh out with the words. If you do not know what this structure is before you start, you will quickly realise that your project is in trouble and you will stop, knowing that it is pointless to continue. Writer's block has set in.

Knowing what structure to use can be hampered by not fully understanding what you are expected to do. Students often have this problem; they do not really know what the teacher wants. Solve this problem by re-reading the instructions. Teachers are careful to put all necessary information into the task description. If you have read and re-read the description and still do not understand, it is time to ask the teacher for clarification. This applies to everyone, not just students. Seek clarification, but make all reasonable efforts to solve your own problem first.

A thesis, for example, has a certain structure; it starts with the research question, then the literature review, then the research approach and data gathering, then the analysis and findings, finishing with the conclusions that can be made based on the evidence. All theses must follow this basic structure. There can be infinite variation of content, but the structure remains.

Finding out what the structure of your document should not be difficult. Your organisation or university/school should have a style guide, or even better a template library for various document types.

If not, the internet is a rich source of free information and templates. Just be sure to verify that what you have found is correct. This means finding two or three examples from different reliable sources (academic, government, library, Google Scholar) and looking for what each of them has in common (the common denominator that they all have). If you base your structure on what is commonly done, you should not go wrong.

Once you have your structure clearly defined, writer's block dissolves away as you see the way forward opening up.

Chunking

The sheer size of a writing task can be daunting. A PhD thesis can be up to a 100,000 words in length (though many will be much less). While most people will never undertake such a mammoth task, the principle is the same with smaller scale but still substantial writing tasks. It is called chunking.

Chunking simply means breaking the task down into smaller, manageable chunks that you can get your head around and deal with. The challenge is to forget that there is still a lot more to do. The thought will nag at you, but you must put it from your mind. Tell yourself, work now, play later. Concentrate fully on completing the chunk that you are currently working on, then move on to the next chunk.

Do not expect perfection

Expecting perfection is another expectation (see earlier section) that needs to be managed. All writers are not created equal. The vast majority of us are not great writers and so we should not expect our work to be of a similar standard to the greats. Instead, aim to be competently good at what you do. If you want to be great, then know that an enormous amount of work lies ahead, but first you must reach the level of being competently good.

Yet this quest for perfection paralyses some writers. The shortfall between their expectations and reality creates writer's block.

As the motivational speaker Denis Waitley observed, the task of writing is more like painting a barn than painting the Mona Lisa. Once you have the nature and scope of the task clearly in mind, you simply get on with the job and work away at it until it is finished. In all likelihood it will not be great art, but rather a competent job. The world will not expect a Mona Lisa, and neither should you.

Once prepared, start writing

While it is important to be prepared, there comes a point where you simply need to start writing. If the words become stuck in your mind like a log jam in a British Columbia river, you need to get the metaphorical dynamite out to break the jam and get things flowing. So simply start writing.

Expecting perfection (see previous section), or expecting the fully formed and finished end result to flow out of you at this point is unrealistic. Lower your expectations and see this stage of the process as a kind of brain dump. Start writing, get the ideas down and worry about polishing them later.

Professional writers do this. There are many stories related by great writers of getting the ideas down in rough form first, then revising and editing them later as a separate activity. Doing it later gives you a degree of objectivity over what you have written.

Start writing, but not necessarily at the beginning

Starting at the start is not always the smartest place to begin.

The opening paragraph needs to do full justice to the rest of the piece, and often you will not know what to say in the first paragraph until you have written at least part of the rest of the piece.

You should have he structure worked out by now, so look at the headings and choose one that you comfortable with and begin with that. Come back to the opening paragraph when the rest of the piece is written.

The concluding paragraphs are also important to the overall impact of the piece. In psychology there is a phenomena known as the Primacy and Recency Effect. It basically means that people remember the first and last thing that they hear/see, so make sure your opening and closing are as good as they can be.

Given the Primacy/Recency Effect, it makes sense to write the middle first, then finish the job by writing the opening and the closing.

Editing & proofing

After the initial draft, discussed in the previous section, comes the editing stage. Do not try to edit when writing the first draft, you will be unable to see what you have just written with the necessary objective eye. Let some time go by before editing. Do something else for a while and come back to it.

There is a well-recognised phenomenon in writing that the writer does not wish to see the mistakes on their own work. We overlook faults in our own work because it is a knock to our self-esteem to see them. The best way is to simply create some psychological space between your creation and the necessary task of editing it. It might feel like killing your own children to change and delete what you have written, but you must not get so attached to it.

Once you have the content as you want it, its time to apply a final coat of polish; the proof reading.

Fear of failure

The fear of failure is one reason people get writer's block, and possibly the main reason why people under-achieve in life generally. Put another way, successful people are prepared to accept the pain of failure as the necessary price for success. No-one succeeds all of the time, but with good planning you can maximise success and minimise failure.

Failure is humiliating and most people will do anything to avoid it. In so doing they sabotage their own success.

If you prepare properly, you will probably not fail at the writing task ahead of you. So put the fear of failure and social humiliation out of your mind. It is a waste of time. On the rare occasions it does occur, resolve to learn from it and move forwards.

Environment

Most people work best in a quiet, comfortable environment, as free as possible from interruptions and distractions. This is easier said than done in many workplaces, particularly when the telephone never stops ringing and co-workers frequently stop by to chat.

It is important to arrange a time and a place during the working day where you can work in a quiet, interruption free environment, since you need to be able to concentrate and follow a train of thought for an extended period. This could be early in the morning or late in the afternoon when few people are about.

Where possible, adjust the lighting to a level that your eyes are comfortable with (not possible in modern, fluoro-lit offices). Reddish or yellow mood lighting can create a sense of security and a relaxed frame of mind in which ideas easily flow.

Routine

Writing is a mental process that can be cultivated with regular practice. Most professional writers have a rule whereby they write a certain number of pages, or write for no less than three hours every day.

The process of writing involves using the part of your mind that performs the enormously complex task of turning ideas into language. Unless you use this acquired skill regularly, it falls into disuse and does not function well. It is similar to physical fitness. Just as regular exercise keeps a person fit, writing something every day helps keep your writing faculties in good working condition.

Schedule a period each day to work on the documentation and do everything you can to stick to the schedule. If your other commitments make it difficult to allocate time on a regular basis, discuss the matter with your manager with a view to reorganising your work schedule. It is not an unreasonable request.

Give yourself a break

This applies where you have just finished a major piece of work and you are emotionally and physically exhausted. You need time for rest and recuperation. Not always easy to do when there are competing demands on your time.

There is something of a fine line to be found here. On the one hand, if you have just finished a big writing job, you may be tired, but you have built up a certain amount of momentum; a valuable commodity for any writer. You are well-advised to not squander this momentum by taking such a long break that you no longer feel like doing any writing.

Take a short break doing something physical, preferably outside in the natural world where you can encounter Nature and experience it. Nature has a wonderful way of restoring your inner balance and building up your energy levels.

The process of technical communication

Much of the material in the previous chapter on writing white papers, reports and university assignments is applicable when writing other technical documents. There are certain aspects though that is specifically relevant to technical writing. This chapter explores these.

To clarify, the previous chapter describes how to write *non-technical* reports and essays. This chapter deals with writing *technical* documents.

Preparation

Preparation is another name for planning. The return on investment for good planning is in the order of 5 to 1, that is, for every one hour you spend planning a project, you save five hours later in fixing defects and re-working something that needed to be done differently. This principle is true for most projects; documentation projects are no different.

Establish your technical writing objectives

You must be clear about what exactly you need to achieve. Technical documents fall into one of three categories:

- Instructive technical procedures
- Descriptive technical information

- General conveyance of technical information for managerial purposes

The three categories are often not separated.

Once you know what category your document falls into, further clarify your objective(s) into specific details. If this is not done, the writing will be vague.

For example, 'To determine the best accounting package for XYZ Industries.' is too general, whereas 'To determine the accounting package which best suits XYZ Industries specific requirements. The choice is between ABC, DFG or HIJ accounting packages.' is more specific.

Identify your readers (audience analysis)

Having a clear understanding of who your readers will be is essential. You do not need to know their names, but you do need to know what age, gender, technical understanding they have. Anything you write must be written with these people in mind, using language that you believe they will best understand.

The following questions can help you clarify who your audience is:

- What does the reader need to know? How deep to go. Is it an in-depth study or a brief, simple memo/report? This determines how much to write, and about what.

- What do the reader(s) already know? Is it about an on-going project that everyone is familiar with, or is it a new project about which little is known?

- Will jargon be understood? Must know and understand the reader to effectively use jargon. If you don't know the reader, use commonly understood and defined statements. For example, do not say 'as yet, there are still several *glitches* in the software'. Better to say 'as yet, the software is still giving out several wrong responses'.

- How much definition of terms will be required? Technical terms must be defined (while jargon is replaced by common words) based on reader's level of understanding.

- Are the primary readers technical?

- If so, are they theoreticians or technicians?

- Are the primary readers managerial?

- If so, keep it general, including essentials.

- Are the primary readers general?

- Can span many levels of understanding.

Determine the scope of the writing

You must know how deep you need to go before beginning writing. Depth is determined by your writing objective and the mix of reader types.

Research

Taking sufficient time and effort to properly research and understand a subject is essential to good outcomes. Not making enough effort in this regard is one of the primary reasons why documents fail to achieve their required objective.

Research by gathering factual information:

- **Primary research** - done by yourself. Includes: experiments mail questionnaires, personal interviews, personal interviews, telephone interviews, personal observation and experience.

- **Secondary research** - done by someone else. Sources include: books, periodicals, newspapers, government documents, industrial and trade directories, and the published results of experiments.

Organise and develop your thinking

Having properly researched a topic and arrived at a clear understanding of it, the next thing to do is think about how best to organise the information. For any given task, there is an optimal way of organising it. Make it a priority to find what that optimal way is.

Here are seven ways you might go about this:

- General to specific development
- Specific to general development
- Chronological development

- Sequential development
- Cause and effect
- Comparison
- Spatial development

General to specific development

Takes a general statement, concept or position and then moves towards the specific elements that support this general position.

Good for developing a highly specific conclusion based on evidence and other supporting material.

Specific to general development

Takes specific statements, concepts or positions and combines them to form a whole.

Good for making inductive or deductive conclusions based on the evidence.

Chronological development

Used where events or procedures must be shown in the time order in which they occur. (i.e. historical account)

Sequence is either past, present, future, or future, present, past.

Sequential development

Order of events determines the structure. The order is determined by the criteria used to order the sequence of events.

Order of importance or necessity are two such criteria.

Cause and effect

Process in which one element causes the next. Can use cause/effect, or effect/cause.

Can deal with either single or multiple cause(s) and effect(s).

Comparison

Shows how things are similar to and different from each other.

Useful in explaining unfamiliar concepts by comparing them with familiar ones.

Need a basis for comparison: cost, strength, reliability, or cost versus reliability for example.

Spatial development

Relates the position and space that the physical elements occupy.

Used to describes machines, buildings and other 3 dimensional objects.

Develop an outline

When you have decided on the most appropriate way of arranging the material, developing an outline is the next task.

The value of outlining

The outline is like a roadmap - it displays the route you will take from the start to the finish of the document. When you take a high-level look at the outline, it must make sense, seem logical, complete.

Outline formats

Outlines can either be:

1.   Simple list. Good for brief, simple documents like memos, letters. Does not allow for expression of complex ideas.

2.   Decimal outline (like this document). Nested headings using decimal notation (eg. 1., 1.1, 2., 2.1 etc.)

The three basic outline styles

1.   **Topical outline**. Uses only topic headings or short phrases.

2.   **Sentence outline**. Each topic uses a complete sentence. This forces you to think in greater detail. It is tempting to string the sentences together to form a choppy prose, though this should be avoided unless there is good reason to use it.

3.   **Paragraph outline**. All main topics are summed up with a complete paragraph. All subordinate entries are then structures as complete sentences.

Do a rough draft

Here is where you actually get down to do some writing. Notice how there have been three earlier steps to complete before starting on this? Many inexperienced writers jump straight into the rough draft stage without doing the preparation and planning. Not performing these makes it much more likely that the document will be sub-standard, or not useful for its purpose.

How to write a rough draft

- Start with your outline, take everything in your head relating to each outline point, and turn the outline into prose.

- Do it quickly, don't wait for inspiration. Just get the words down, regardless of how bad it seems. You can fix it later. It is important to keep the flow of words going.

- Resist the temptation to correct typos and mistakes (same reason as previous point).

The rough draft cures writer's block

- Rough draft is major cure of writer's block.

- Outline tells you what to say. Rough draft puts flesh on the bones.

- Don't show anyone the first draft so that no one but you will know how bad it is (if it *is* bad).

Revise

Revision converts raw creative prose into flowing, readable writing.

Activate the language

Active writing creates interest and understanding. Verbs are central to active writing.

Check active and passive voice. There is usually a lot of passive voice in rough draft (eg. "I heard it through the grapevine" is active voice, "it was heard through the grapevine" is passive.

Another example; stress is exerted by the load pushing down on the platform (passive). The load exerts stress on the platform by pushing down on it (active).

First, the electrical energy input of the motor must be calculated (passive). First you must calculate the electrical energy input of the motor (active)

Activate suppressed verbs. Suppressed = weak. They stop the verb from being more effective. This is usually the suppression of the active verb.

Words ending in '-tion', '-ing', '-ent' and the use of prepositional phrases with verb forms.

For example, completion of the heat tests has been accomplished by the lab (completion is suppressed). The lab has completed the heat tests (activated).

The software suite is dependant on a dedicated server for its operation (operation is suppressed). A dedicated server operates the software (activated).

Clarify the writing

If the reader does not understand, your writing has failed. When your writing is clear, the reader can go directly to your ideas without struggling with the writing.

Check the following:

1. **Decide on personal vs. impersonal reference.** Will you use 'I' and 'You' (personal). Impersonal tends towards the passive. Use personal unless required to use impersonal.

2. **Use connotation and denotation.** Dictionaries denote meaning. Emotive words have connotative meaning. Connotative words are far more easily misunderstood. Waste/junk, miscellany/odds and ends, things/stuff, offspring/kids, man/bloke, woman/sheila, compact/smash.

3. **Determine whether jargon is helpful.** Jargon is exclusive language because it is coded to exclude its denotative meaning. Boot up, bit, byte, megs, bug, I/O, ports, CPU, RAM, ROM.

4. **Change abstract words to concrete words.** Vague abstract words obscure meaning. Can mean anything to anyone, or be imprecise in the context. The supports must be strong enough (abstract). The supports must have a tensile strength of ... (concrete).

5. **Eliminate affected language.** Uses complex and impressive vocabulary that says little. Tries to sound official, legal, and scientific.

Admonish/warn, altercation/dispute, anathema/curse, approximately/about, ascertain/find out.

6. **Replace clichés and trite language**. Clichés interfere with clarity and annoys the reader. Elongated word forms, superficial use of foreign words, and idiomatic expressions all detract from clarity. At this point in time/now, consensus of opinion/agreement, during the course of/during, in the vast majority of cases/in most cases, on a weekly basis/weekly, refer back to/refer to, until such time as/until, due to the fact that/due to.

7. **Correct misplaced modifiers**. Modifiers (gerunds, participles, infinitives) can help with clarity. When incorrectly used, they confuse. (a) the topic of the seminar was digital conversion of analogue systems at our Sydney office, (b) the topic of the seminar at the Sydney office was digital conversion of analogue systems. Both are plausible, but only the writer knows which is correct.

Simplify the writing

There is a big difference between *simplified* writing and *simple-minded* writing. Don't make the reader work harder than is necessary to get the information.

1. **Keep sentence length down**. Longer the sentence, the harder it is to read. Becomes a real problem when a document is made up of many

long sentences (>30 words). Short sentences combine to make for clearer meaning. Aim for 12 to 15 words per sentence.

2. **Keep word length down.** Also makes it easier to get past the writing to the meaning. Technical terms may oblige the use of long words.

3. **Eliminate needless words.** Shipping considerations will be easy because of the flexible nature of the materials. Shipping will be easy because the materials are flexible.

4. **Simplify positive and negative constructions.** Present positive and negative statements in their most simplified construction. Were not a success versus failed.

5. **Watch for the 'It ... that' syndrome.** It has been shown that, It can be proven that, It is a known fact that.

Do the final draft

The final draft is where you evaluate the usability of the document. Ask yourself, *if I were the reader, would this work for me?* Try to take yourself out of your own self (the person who would like this job over and done with) and the put yourself into the position of the typical reader (as determined by your audience analysis).

Allow for generous use of white space

Solid blocks of small text are very discouraging, not to say confusing in many cases. Readers want to feel that a page can be easily read.

You might be surprised at how little extra space is used when you include white space in a document. Try this experiment sometime. Take a document that has the Body Text (or Normal paragraph style) set to zero indent. Note down the number of pages with these settings (make it at least 20 pages long for best illustrative effect). Then redefine the Body Text/Normal paragraph style to be 3 cm. Apply the change and note how many extra pages are included. It might only be two or three. Sure, you will have saved a few pages by eliminating whitespace, but you will have made your document considerably more difficult to read.

Always consider your reader and what will make it easier for them to extract the meaning from your document.

Use topic headings often

- Topic headings open up the text and produce white space. Readers are discouraged by unbroken text. Headings create manageable pieces.

- Headings indicate what is coming next.

- Headings alert the reader to major breaks in the writing and its concepts.

- Headings allow reading out of sequence.

Use listings

- Lists save time.
- Lists make it easier to see the various elements of the list.
- Use ordered list when order is important, otherwise use bulleted list.

Use illustrations effectively

A picture is worth a thousand words, provided the picture is appropriate.

Line drawings, tables, pie charts, bar graphs, line graphs, flow charts and logic diagrams, schematics, photographs, cutaway diagrams, exploded diagrams.

Include adequate appendices

Use appendixes at the end to supplement and/or clarify.

Appendixes contain relevant material that is either not critical or is too large to place in the main body.

Do the final presentation

The document should be properly bound because it looks professional, and because it the pages from getting dirty or falling out.

Project management issues

In larger scale projects, it may be necessary for project management to be performed. This section outlines several issues relating to this.

Collaborating

Some documentation projects require the input of more than one person. A document is prepared by several people, each completing one or more section, then the pieces are assembled.

The important principle to observe with collaboration is to have the most experienced and/or knowledgeable person act as project manager who coordinates the efforts of the team members.

The project manager takes particular interest in the activities early on in the project, particularly those relating to categorising the document and deciding how to structure it.

The Software Development Lifecycle (SDLC)

The Software Development Life Cycle (SDLC) is the overall name given to the complete process of developing software, from the earliest planning through to maintenance and eventual retirement.

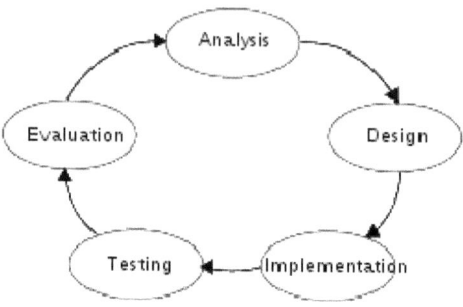

It includes requirements, validation, training, and user (stakeholder) ownership. There is technical documentation required at all stages of the SLDC.

Software internationalization & localization

Internationalization and localization is the process of adapting software to different languages, regional differences and technical requirements.

Internationalization is making the software readily able to various languages and regions without engineering changes.

Localization is about inserting the local language of the target region into the internationalized software code by adding locale-specific components and translated text.

The terms are frequently abbreviated to *i18n* (where 18 stands for the number of letters between the first i and last n in internationalization, and *L10n* for localization.

A technical writer is often involved in the localization of software.

Persuading an audience with effective argument

You could have the most brilliant and creative mind of your generation, but unless you have at least a basic understanding of how to persuade a sceptical audience, it is unlikely that your ideas will see the light of day.

Sir Isaac Newton, for example, worked for many years in relative obscurity producing his most brilliant work, but it was not until Sir Edmund Halley (of comet fame) persuaded Newton to present his ideas to the world that he achieved well-deserved recognition.

Newton would probably have fitted in to a modern IT development environment quite well, where people tend to be introverts who prefer to interact with technology than with humans. Team work is discussed in a later chapter.

This chapter outlines the basic principles of persuasion. They can be applied in almost any situation.

Persuading others to your point of view

Argumentation uses clear thinking and logic to convince readers on a controversial issue.

Persuasion is involved if appeals (ethical, logical, emotional) are made in the process.

Argumentation and persuasion are often combined because people react rationally and emotionally to situations.

How argumentation-persuasion fits purpose & audience

Argumentation and persuasion are everywhere, in TV, radio and newspaper advertisements and editorials.

- The *logical appeal (logos)* relies on sound logic with supporting evidence that is unified, specific, sufficient, accurate, and representative.

- The *emotional appeal (pathos)* is directed at the audience's needs, values and attitudes. It usually encourages them to commit to a course of action. Connotative words are the tools of this appeal.

- The *ethical appeal (ethos)* relies on the writer's credibility or reliability. People won't believe you if they think you don't know what your talking about.

Audience analysis helps you to determine the best appeal, or combination of appeals.

Audiences can be categorised as follows:

- **Supportive audience.** If the audience already trusts your credibility and agrees with your position, an emotional appeal can be the primary appeal.

- **Wavering audience.** People who are interested in your argument, but don't necessarily believe you. They may not be as informed on the subject as they should be. *Ethos* and *logos* work best here.

- **Hostile audience.** An apathetic, sceptical or hostile audience is obviously the most difficult to convince. Use *logos* primarily. Avoid *pathos*.

Strategies for using argumentation-persuasion

1. Identify controversy & define your position

At the beginning of the paper, identify the controversy surrounding the issue and define your position. This allows the reader to know immediately what your position is.

For example, Australia's social security system has been abused over the years. (broad thesis)

No-one except the handicapped and mothers of pre-school-age children should be eligible for benefits.

2. Provide strong support for your thesis

Supporting evidence *must* be (a) unified, (b) specific, (c) sufficient, (d) accurate, and (e) representative. The evidence can be based on personal experience, or from external sources; statistics, facts, examples, expert opinion taken from books, journals, reports, interviews, and documentaries.

If you uncover information that undermines your thesis, do not ignore it, use it to arrive at a more balanced viewpoint. Your opponent will probably use it, and if you can pre-empt what s/he says, you will be more persuasive.

3. Avoid using a hostile tone

Audience is alienated by hostile or condescending language. *'Anyone (with half a brain) can see that ...'* or *'Its obvious even to a moron like you ...'*

Also avoid personalising the argument. *'My opponents find the law ineffectual'* sounds adversarial. *'Opponents of the law find it ineffectual'* sounds more even-handed.

The even-handed approach helps to establish common ground. This makes you appear fair and reasonable and therefore believable.

4. Organise the supporting evidence

Emphatic approach. The most important or compelling evidence is saved for last. Has built-in momentum because it starts with the least important and builds to the most significant. People remember best what they experience last.

Simple to complex approach. Proceed from simple concepts first to the most complex last. Gains audience agreement and rapport. It assures them that your argument-persuasion is firmly based in shared experience. Conversely, you risk alienating an audience if you start with difficult technical and/or complex material.

5. Acknowledge and refute differing viewpoints

Good argument-persuasion seeks out differing viewpoints, acknowledges them, perhaps even admits that they have some merit. This strategy strengthens your argument by:

- Anticipating objections,
- Highlighting flaws in your own position,
- Making you appear reasonable and thorough.

Three techniques for dealing with dissenting positions:

- Two part proposition that identifies the opposing opinion and states your position, implying that your position stands on more solid ground.

- Use one or two paragraphs to summarise the opposing argument, granting where appropriate the validity of some of its points. Then present the evidence of your own position.

- Refute all or part of the dissenting view by pointing out its flaws.

6. Use induction or deduction

Your line of reasoning is a sure indicator of how rigorous you have thought through your position.

Inductive reasoning. Examines specific cases, facts or examples. Based on these specifics, you then draw a conclusion or make a generalisation. Used by scientists, eg. `smoking increases the risk of cancer'. The evidence used must be accurate, recent, and representative if your case is not to be undermined.

Limitations of induction is that there is a significant level of uncertainty. The conclusion can never be more than **inference,** involving an `inductive leap'. There could be more than one explanation for the evidence cited.

Deductive reasoning. Begins with a generalisation that is then applied to a specific case. This move from the general to the specific involves a three-step process - the **syllogism:**

- Major premise - a general statement about an entire group.

- Minor premise - a statement about an individual within that group.

- Conclusion - a conclusion about that individual.

For example, `In an accident, bigger cars are safer than smaller cars (major premise). The Ford Falcon is a large car (minor premise). In an accident, the Falcon will be safer than a smaller car (conclusion).

A syllogism is valid if the major premise is not a sweeping generalisation, or the minor premise is not faulty in some way.

7. Toulmin logic to connect evidence and thesis

Whether you use induction or deduction, your argument depends on strong evidence. Stephen Toulmin proposes this approach to strengthen the connection between evidence and thesis:

- **Claim**. The thesis, proposition or conclusion.

- **Data.** The evidence (facts, statistics, examples, observations, expert opinion) used to convince the audience.

- **Warrant.** The underlying assumption that justifies moving from the evidence to the claim.

For example; `The train driver was drunk when the train crashed' (data). Transport workers entrusted with public safety should be tested for drug use (claim). Transport workers should not be allowed on the job if they use drugs (warrant).

Ask yourself:

- What evidence do I need to convince the audience.

- Is my warrant clear? Should I state it explicitly or implicitly? What backup do I have to justify my warrant?

- Would qualifying my claim make my argument more convincing?

8. Recognise logical fallacies

Ad hominem *fallacy*

To the man - occurs when someone attacks a person rather than their point of view. A poor substitute for reasoned argument. For example, attempting to destroy the credibility of an opponent by saying they have a drinking problem or a criminal record etc.

Appeals to questionable or faulty authority

Most people are wary of statements like `sources close to ...`, `an unidentified spokesperson says ...`, `Experts claim ...`, `Studies show ...` If a source is reliable, it should be named. Sources hiding behind anonymity can say anything and get away with it.

Begging the question

Failure to establish proof for a debating point. The writer expects the audience to accept as given a premise that is actually controversial.

For example, *if you say the Governor General should be stood down because his appointment is unconstitutional, you would have to first prove how his appointment is unconstitutional.*

False analogy

Wrongly implies that because two things share some characteristics, they are therefore alike in *every* respect.

For example, *if you compare tobacco with cannabis, saying both involve health risks and are addictive, then conclude that because smoking tobacco while driving is legal, it should also be legal to smoke cannabis while driving.*

This overlooks a major difference between the two - that cannabis is known to affect co-ordination and perception, while there is no evidence that tobacco does the same.

Either/or fallacy

Where it is assumed that a particular viewpoint or course of action can have only one of two diametrically opposed outcomes.

For example, `Unless universities continue to offer scholarships based solely on financial need, no one who is underprivileged will be able to attend university.'

This ignores the fact that some bright but underprivileged students might get a scholarship based solely on their academic excellence.

Red herring

An intentional digression from the real issue - a ploy to deflect attention from the issue.

For example, `Mass killing of wild goats in western Queensland is immoral'. If you reply, `The goat is an introduced species and shouldn't be there' is a red herring.

Persuading international readers

Some adjustment may be necessary when addressing international readers.

What is considered to be a benefit differs from country to country. *For example, in Australia, beer is served cold. In England beer is served at room temperature. Ads for Australian beer in England (XXXX btand) focus instead on its thirst quenching qualities, i.e. `Australians wouldn't give a XXXX for anything else', showing a road petering out at a Hotel somewhere in the*

Australian Outback. The road-builders had simply given up when they reached the Hotel, presumably because (a) they were so hot and tired, and (b) XXXX is so good.

What readers consider to be good reasons differ from country to country. In Western countries, decisions are generally made based on fact _ or should be; whereas in Middle Eastern countries, the facts are used to support a primarily emotional appeal.

What is considered to an appropriate role differs. In strongly hierarchical societies, the writer must avoid appearing disrespectful, whereas in Australia, a writer can afford to be more direct at the risk of offending superiors.

It is important to be fully aware of the cultural norms of the target audience.

The ethics of persuasion

Persuasion should *never* be used to further one's own interests at the expense of other people's interests.

Even if it succeeds at the time, *unethical* persuasion is usually revealed as such at a later time. The deceived audience is likely never to trust the writer again, and may actively warn others not to trust the writer.

Do not mislead

Your reader has a right to evaluate a situation based on accurate facts and figures, as supplied by you. You must not intentionally supply misleading information, even by only leaving certain information out, so that the reader is basing their decision on a false premise.

For example, *if a salesman tries to persuade you to buy a Brand A computer instead of a Brand B computer because the chip in B is known to be faulty, he is leaving out the important fact that the fault only appears when a calculation to more than 64 decimal places is being performed.*

Do not manipulate

German philosopher Immanuel Kant said that we should (a) not do something that would not be acceptable if everyone did it, and (b) not to dehumanise people, or treat them as a means to an end. These are known as Kant's first and second categorical imperatives.

The action we are trying to persuade others to take must advance *their* interests, as well as our own (Win/Win situation).

High pressure sales techniques are often unethical.

Be open to your reader's point of view

Empathise with the audience to determine their real needs. If they have a legitimate objection, based on their real needs, then you need to modify your position accordingly, rather than see it as an objection to be overcome.

The reader is not an adversary; they are a partner in your joint search for a course of action that is suitable to everyone.

Argue from human values

If human values are relevant, always mention them. If they are disregarded initially, they are likely to be recognised at a later point.

For example, when trying to persuade an organisation to purchase a new computer system, it could be pointed out that such a system would improve the quality of the staff's working conditions by freeing them from boring, repetitive jobs, allowing them to focus on the kinds of jobs help them to develop.

Writing user documentation

Introduction

User documentation is basically about finding the best way of putting across technical information in a non-technical way to non-technical people. If we want to do the greatest amount of good for the largest number of computer users, then it is the end-user documentation audience who require the most attention, since they are the largest consumers of computer documentation

User documentation[i] is the most difficult type of communication for IT people (*"techos"*) to do well. This is because user documentation calls for communication between people with widely different backgrounds.

User documentation does not necessarily require writing that simplifies the software. User documentation is best thought of as a type of writing that *translates* the computer activities for the benefit of non-technical users and readers. User documentation therefore takes data processing information and translates into ideas that can be readily understood by people who are skilled in other areas or other disciplines.

User documentation presents a situation much like that involved with translating a foreign language. When you are speaking to someone who shares a common background and language, they can fill in the gaps and make up for mistakes in your communication. However when speaking to someone who has a different background and speaks a different language, they cannot make up for gaps and mistakes, and need additional explanation.

User manuals are not technical

Think of the difference between a programmer's reference guide and the user guide that comes with common software packages. The first contains highly technical language and concepts, while the second is expressed in the kind of language the user is likely to understand.

Whereas technical writing catalogues facts in detail, user documentation should only say as much as is necessary to describe a step by step process. Too often, a document intended for user consumption is confused with technical writing, resulting in a document that is doomed to failure unless the user is willing and able to think like a technician.

Why we need good user documentation

Without a doubt the computer industry suffers from a lack of good user documentation and it is the user who suffers. More often than not the software is written and then someone says, "Oh, and of course we'll need a manual". So, taking the technical specifications that the programmers used, someone puts it through a word processor and it comes out the other side as the "user manual". The problem is that it still sounds like a technical document. The user takes one look at it, thinks "This is useless" and after that the manual stays on the shelf gathering dust until it falls victim to a cleanup some years later.

On the other hand, large software companies spend a lot on their manuals in order for them to be "friendly" to the average user. That is to say, presented so that the user can easily find the information they need to get on with the job at hand.

The same principles which the big software companies use to produce those manuals are given in this book. If you follow the steps outlined in this guide, your manual will be useful to the reader.

Manuals written in the early stages

As far back as the 1980's the then powerful Digital Equipment Corporation (DEC) noted in their internal documentation guidelines that user documentation should be written first - not last as is traditionally done - because the user documentation is an excellent way to debug the design of a system or program. If a writer finds it difficult to document a system, the problem is probably the system, not the writer. Holes in design, obscure constructions and apparent contradictions become starkly visible in the documentation.

Potential problem areas can be identified early rather than after the product is finished. If problems are identified early, they are more likely to be fixed, since fixing problems later in the development cycle costs more money. In fact the later a problem is found, the more costly it becomes to fix.

Most software products that can be described as being outstandingly good and/or user friendly have been developed this way.

On-line manuals / help files

There are a range of proprietary on-line manual / help file authoring tools available on the market if the reader needs them. The quality and functionality of these tools varies. With the changing landscape of this market, this book makes no recommendations. You can choose the package that best meets your needs, based on your requirements.

But even if you will be only be producing on-line and not paper-based manuals, you still have to know *what* to put into your online manual, and *how* to organise the information for best effect. The information in this book is still useful for this.

So if you are developing on-line help, use this guide to write the manual, the content and organisation of which can then be fed into the help authoring tool, and/or printed as a paper-based manual.

Preparation

This chapter covers the range of activities that are necessary to establish a good foundation for your documentation project. The importance of good preparation can hardly be emphasised enough - it is vital to the success of any documentation project, particularly larger projects.

Preparation is another name for planning, and as the old saying goes, *failing to plan is planning to fail.* The Return on Investment (ROI) in software development projects generally has been estimated at around 6:1. For every hour spent planning, six hours are saved or gained later in the project.

The object is to clearly define the finished product before beginning to write. If you do this, and get approval for your concept so that the project manager and yourself both expect the same thing, you will avoid the need to re-do sections of the manual later.

Much trouble can be avoided by having a clear agreement between yourself and the project manager as to what the finished product will look like and contain. Unfounded assumptions must be avoided since they are usually not shared by the parties.

In a sense, good planning is about creating a shared vision of what the future product, in this case a manual, will look like. As every project manager knows, one of the hardest things about running projects is managing stakeholder expectations. Good planning does just that.

Source material

It is important that all source material necessary for a thorough treatment of the subject be made available to you at the outset.

Source material includes the following:

- All relevant specifications, record formats and screen and report layouts.

- An operating copy of the software, if available.]

- The analysts and programmers of the software, including the timely resolution of questions raised by you.

- Where available, typical users for audience analysis and usability testing.

While it is your responsibility to communicate accurately the source material to the user, you are not personally responsible for the accuracy of the original source material.

Where documentation standards already exist, you need to be supplied with copies of the relevant standards. Otherwise, use the standard outlined in a later section of this book.

Source material received by you is normally on a loan basis. You need to keep it in good order and return it when finished. In some cases the information might be confidential. In these cases, you will need to sign a non-disclosure agreement.

Documentation project plan

It is necessary to develop a project plan. This is done in two stages. The first stage is to be briefed by the project manager or someone delegated by them as to the exact nature of the job. The results of the briefing are a mutual understanding of what is to be delivered, and by when.

Prepare a document called *"Preliminary Project Plan"* using the headings which follow. This will form the basis of the final project plan which can't be developed until the following information is gathered and considered.

- **Title -** working title, purpose, scope and limitations of document.

- **Audience** - who will read the manual?

- **Why used** - reasons why the documentation would be used by the intended audience, and for what purpose.

- **Table of contents** - draft table of contents.

- **Deliverables** - how many printed copies (whether on CD/DVD or printed) are to be supplied, disk and file formats (including software versions) and where and when they will be delivered.

- **People -** who and what resources will be available to help you? What person hours are required? At what cost?

- **Resources -** project resource requirements, including the information and other resources that the customer will have to provide and when.

- **Change control** - a specified method for passing information about software changes to the documenter during the development of the manual.

- **Milestones** - a schedule showing appropriate milestones, including where appropriate:

- Documentation plan approval.

- Preparation, review and approval of each draft.

- Usability testing.

- Camera-ready artwork preparation.

- Printing, binding and distribution.

- **Source material** - what written information - i.e. requirements list, specifications, reports, etc. is available?

- **Software** - is the software to be documented available for you to use?

- **Standards/specifications** - is the documentation to be written to a particular standard or specification? - for example a company style sheet or standard format. The "Document production" section of this manual is an example.

- **Technical edit** - who will check the technical accuracy of the manual?

- **Editorial control** - who has editorial control?

- **Budget** - how much will it cost?

- **Copyright** - ownership of copyright and any other proprietary rights. Usually, the author of a commissioned document is the first owner unless a specific agreement otherwise. If the customer is to own the copyright, it must be assigned as part of the contract between author and customer.

- **Translation** - provision for the translation into other languages, if applicable.

Getting answers to these questions allows you to develop a suitable project plan.

When you have finished the preliminary project plan, discuss it with the project manager and after any amendments which might arise from those discussions, it should be signed-off or otherwise formally agreed by both parties. This might seem an unnecessary formality however from experience its value as a benchmark for future reference is undoubted. It also obliges the project manager to provide you with everything you need to complete the project - this sometimes comes in very handy.

Project plan

Once the preliminary project plan is complete and it has been signed-off, a detailed project plan is developed.

Using the preliminary plan as a basis, develop the project plan using the following headings shown below.

Note: a series of forms to support the following activities are provided in appendix A.

Document name (section 1)

Decide on a name for the manual. Be guided by the subject matter and the intended audience. Be as concise as possible when describing the purpose. A manual can also be called a handbook or a guide.

For example, "CMS User Guide" is more concise than "Complete Guide to the Construction Management System".

Purpose (section 2)

A clear statement of the purpose of the document. This will start with what you put in the preliminary plan, together with any amendments which might have arisen from your discussions with the project manager.

Specify the usage mode of the document. The usage mode of any user documentation can be divided into two broad categories - instructional and reference:

- **Instructional** - where the user needs to learn about the software. Instructional documents are either informational (i.e. tutorials and introductory manuals) or task-oriented (i.e. quick reference manuals which give the steps the user needs to perform a task).

- **Reference** - where the user needs to refer to information or refresh their memory and which give the information needed to understand the subject.

It's important to decide the usage mode at the outset. Without this definition, the manual will probably be a mixture, alternating between modes, which will confuse the reader. On the other hand, the impression of *consistency* inspires confidence in the document - something you very much want if the manual is to succeed.

Instructional mode

An instructional mode document should give the background information needed to understand the system, and also give details of the range of functions the software can perform and how to perform them. Examples are given to reinforce the learning process.

The Informational type of instructional document provides background and any technical detail needed to understand the system. This would include a system overview, a theory of operation and a tutorial. The technical information should be provided by the people who developed the system.

Task oriented instructional documents outline the procedures of operation. Examples of this kind of documentation include a diagnostic procedures manual (i.e. troubleshooting manual), operations manual and software installation manual (getting started or read this first).

Reference mode

Reference mode documents organises information that users might need, and allows quick access to details of specific subjects. Examples of reference mode documents include a command manual, error messages, program calls, quick reference guide, software tools and utilities.

Table of contents (section 3)

Outline the draft table of contents.

Deliverables (section 4)

Specify how many printed copies are required, whether disk copies are to be supplied, the disk and file formats (including software versions) and where and when they will be delivered.

Resources (section 5)

Specify the project resource requirements, including what information is required from who and when. Who and what resources are to be made available?

What source material, written information is to be made available. This includes requirements list, specifications, reports etc.

Nature of the software (section 6)

What is the nature of the software? Is a working copy available for you to use? For example is it for accounting, inventory, and personnel or perhaps to control a production process? Define the nature of the software in terms of it's functions. What is it *for* exactly? Then consider the kind of user interface it employs and the type of work that users will perform with it.

Scope (section 7)

What is the scope of the project? Are a complete set of manuals required which detail every aspect of the software, or would it make more sense to document only those parts which people will use? Consider the benefits against the time and effort that will be needed.

The documentation needed depends on the nature of the software product, how it is applied and who will be using it. Once you've identified these, the document sets for the intended users can then be determined.

Topic (section 8)

Define the topic to be covered. What exactly do the users need to know? Do we want to provide background information to help them understand what's happening behind the scenes, or just tell them enough to do the job? Cost can be a major consideration here.

Who is the audience? (section 9)

Think about exactly who will be using the documentation. This tells you what assumptions can be safely made about how much the user already knows, and the kind of language they will understand. For example a manual which will be used by accounting staff will call for different language than a manual used by station staff.

Think about the different ways that the user will interact with the software. Will they need to interact a lot, and at what level. For example, does the software require fairly simple responses or will it need whole screens full of information to be entered? The answers to these questions determine the presentation style and the amount of detail.

In some cases, it may not be worth producing full documentation. For example a handful of highly technical users might need a relatively short manual so it would not be worthwhile producing a very detailed set of manuals in this case. On the other hand it would be worthwhile for non technical users, particularly where there are more than a handful of users.

Careful consideration of the nature of the audience allows you to choose the following:

- The best language to use.
- The right level of difficulty - not too difficult.
- How much material to include (so as not overwhelming the reader with too much or starving them with too little information).

- How to organise the concepts so that readers progress logically, beginning with what they understand and proceeding into what they don't understand.

When the manual will be widely distributed, its worthwhile keeping to the so-called "lowest common denominator" i.e. the person with the least amount of knowledge or expertise. Following this principle makes sure your manual reaches all the people it can.

Document set (section 10)

A set can be a single document or several documents, depending on how much detail is to be included and the needs of the audience. For example, a set of user documentation might be made up of a two volume reference manual and a training manual. The training manual is clearly separate from the reference guides which have been broken into two since to have one large volume would be unwieldy.

Where a document set needs to cater to widely differing needs, it's necessary to either include different sections for the specific audiences with that audience being clearly mentioned in the introduction, or to simply produce different documentation sets for the specific audience.

Presentation (section 11)

The layout and design of the document is outlined in detail in the later section "Document production". Is the document to be written according to any standard or specification?

Project schedule (section 12)

Develop a schedule showing the milestones, as follows:

- Blueprint.
- Research.
- First Draft.
- Language edit.
- Technical edit.
- Beta test.
- Final review date. Master copy ready.
- Manuals ready for delivery.

Software (section 13)

Indicate the software tools you propose to use for the project. As a minimum requirement, the various packages need to be compatible with each other. Some graphics packages, for example, won't deliver useable images to some word processors.

To avoid potential problems, stay with tried and proven software solutions or seek the opinion of knowledgeable people whose judgement you respect.

Editing (section 14)

Specify who will retain editorial control over the document.

Specify who will perform the technical edit.

Budget (section 15)

How much will the documentation project cost, or how much is being allocated for the documentation.

Copyright (section 16)

Who has ownership of copyright and any other proprietary rights?

In the normal course of events, the author owns the copyright. If the author has been commissioned by a second party, the formal agreement between the two parties usually prescribes that ownership remains with the commissioning party unless otherwise specified.

Translation (section 17)

What provision is to be made for the translation into other languages, if applicable. This includes any Localisation tools that may be used.

Estimating

Estimating the time and resources needed to successfully complete a documentation project can be a difficult procedure for the inexperienced writer. It is easy to underestimate. Estimates can be overrun if the software changes during the course of writing the manual.

The *Minutes and Hours* estimating method outlined below works on the assumption that it takes around three hours per page to write text to publication standard. The time needed to design graphics is determined by their complexity and the number of redrafts needed to ensure their technical accuracy. On average, the kind of graphic commonly found in user documentation will need three to five hours to design and amend. Graphics in this sense do not include screen images.

Knowing how long each page is likely to take does not help you to determine the likely number of pages. This needs to be done using a combination of common sense and a shrewd appraisal of the number of software functions to be documented. Your estimated page-count should be reviewed a fortnight or a month into the project.

When undertaking a large project, the deliverables should be split into manageable parts. The estimated time to complete the entire project would then be given in terms of whole months, with the first part being the only part worked out in detail.

The writing process

This is an important section for people who are not experienced writers of non-technical prose. By applying the principles outlined here, you will extend your writing capability towards a professional level.

You may have already attained a level of competency with writing technical prose, but this is not what is required for a good user manual. User manuals describe technical matters in plain English.

Clear & effective communication

A subject as large as this could fill a library, but as big a subject as it is, there are some distilled principles which can be applied to help you write more readable user manuals.

Take the following guidelines to heart and practice them. They were largely put forward by the English writer George Orwell in his book *Politics and the English Language* (1947). Taught in university courses, practised by experienced writers everywhere, they can be considered some of the most reliable "tricks of the trade".

These principles can help you to use language as an "instrument of expressing and not for concealing or preventing thought", as Orwell said. They apply to most kinds of writing, including user manuals:

No tired figures of speech

It is a fact that when communicating, people often use expressions that have become overworked and tired (i.e. a cliché). The expression may have once had great impact, grabbing the reader's attention with the freshness of their imagery, but after 1,000 uses, they are past their "use by" date and should be retired. Take the time to think of new ways to express ideas and your writing will benefit.

As an exercise, resolve to vigilantly listen to what comes out of your mouth (and what you write) in an average day. You might be surprised at how much of it is cliché. Make a conscious effort to think of new ways of expressing yourself.

Your listeners will in all likelihood appreciate the effort. You will seem more interesting to people.

Short not long words

Never use a long word when a short one will do; use "timely" not "auspicious" or "opportune. Use "set" rather than "predetermined".

It is often tempting to use a large word because we can, because we know it. We would like people to see how good our vocabulary is. Resist this temptation and use the short word. You will reach the maximum number of people with this practice.

Short words tend to be more specific or concrete, making the message more definite. Short words also usually have more impact.

Use a specific, concrete word instead of a general, abstract one. Instead of:

"We should request management to do something about their high overheads", say

"Let's ask John, Susan and Peter to suggest five ways of cutting departmental costs".

Examples of general (usually long) versus specific (mostly short) include:

- stringed instrument/guitar.
- transport vehicle/car.
- public service department/Queensland Transport.
- entertainment/movie.

- science/biology.

- sporting event/Olympic Games.

Specific words help by allowing the other person to see a clear meaning, general or abstract words tend to obscure meaning.

Economical & precise with words

Economical   if it's possible to cut a word out without losing the meaning, always cut it out. For example to write:

"You can begin to download the data to the hard disk of the computer by loading the USB memory stick and selecting "Download" from the Utilities menu which is found in the System Administration area.."

Is not as economical as:

"To download the data, insert the USB stick and select "Download" from the Utilities Menu".

They both get the same meaning across but the first includes extra words which add nothing to the clarity of the statement, but which the reader is obliged to plough through nevertheless. In this example it isn't necessary to tell the reader where the downloaded data will go or where to insert the USB stick or even that the Utilities menu is in the System Administration area if this section is dealing with the System Administration area as a whole.

The rule of thumb is, don't make people read more than they need, the extra words get in the way, waste time and cause irritation.

Precise   With around 500,000 words, English has perhaps the largest number of words of any language. With such variety, try to choose the words which best express your thought while bearing mind the advice on keeping it simple. Many words have only slight differences in meaning; i.e. *assisted, benefited, served, helped.* Or *meritorious, illustrious, distinguished, significant, renowned.*

The best way to achieve precision is to:

- Think carefully about what you're saying, and
- Have a broad enough vocabulary or use a Thesaurus. A good way to build your vocabulary is to make a point of looking up words you don't know and perhaps using a thesaurus when writing a document.

Active not passive

Always use the active voice where possible. Active voice has more impact than passive voice and is usually more concise as well. For example it's better to write:

'use the active voice'

than it is to say:

'the use of passive voice is to be discouraged'.

Notice the diluted effect that the passive voice creates. An enormous amount of what is written in organisations suffers from this problem. Why? It is partly through habit (pre-fab expressions), partly through a desire to lend an air of authority to the words and partly to hide a lack of real understanding of the subject, or worse to conceal the real meaning. Half-baked or incomplete thoughts tend to be expressed this way.

Everyday English not foreign, jargon or scientific

Except in situations where these are specifically called for, everyday English should be used rather than foreign, jargon or scientific words (i.e. not used for the sake of appearing knowledgeable). As a general guide, choose words that are likely to be understood by the largest number of people unless you are writing for a highly specialised readership.

It is often more difficult to use a common word when the concept is normally described in technical terms. Never assume that people know the meaning of technical words unless they have specific training (i.e. a computer science graduate can be expected to know computer jargon, but the accounts clerk who is actually using the software cannot be expected to understand computer jargon.

Prefabricated language

Orwell also pointed to the habit many people have of using "prefabricated" language. Rather than making the effort to think of new ways of describing things, most people lazily continue to use the same old expressions they've been using for years. For example:

'At this point, the weekly invoice run is initiated and without further ado will run until finished.'

Contains two pieces of prefabricated language; *"at this point"* and *"without further ado'*.

The result of overused expressions is that the message may not get through since the reader has tuned out after encountering too many overworked phrases. Original sounding language helps get the message across by sparking the reader's interest. In the above example, you could say:

'The weekly invoice run now commences."

Not using prefabricated language also leads to the economical expression of ideas.

Present tense not past/future

Unless it specifically applies, use present tense. Say *"Pressing <enter> accepts the default value"* rather than *"Pressing <enter> will accept . ."* (future tense). Another example, *"use active voice in the present tense"* rather than *"the use of passive voice in the future tense is to be discouraged'.*

Using present tense makes the message sound more immediate. The reader unconsciously thinks if it's happening now, it is worth knowing. If it's happening in the future, let's wait until it happens. If it's already happened, it's history.

Avoiding overstatement

This general guideline applies to all communication. While there are few opportunities for overstatement in user manuals, it's still worth mentioning. In an attempt to strengthen their message, many people resort to overstatement - words that convey an exaggerated view of a person, event or situation. If someone says "You never help me with my work" they invite a reply like "Of course I help you, what about last week?'

When a speaker exaggerates it usually makes the other person defensive - all of which gets in the way of clear communication. It's better to limit yourself to simply stating the facts; it shows that you're being fair and mindful of the other person's feelings.

Adapting words to the reader

To help the other person perceive what you are saying as interesting and intelligible. Certainly, using precise specific words adds interest as mentioned earlier, but you can also add interest by being concise and colourful in your phrasing.

Another way to add interest is to use colourful, non cliché expressions. For example, to describe an experience as being "electrifying" is colourful but commonplace, to say it was "like touching an electric fence" adds colour and freshness, making it both more interesting and entertaining for the listener/reader.

Never barbarous (advisory only)

Note: This section is for general interest only. It is included to complete Orwell's excellent list. Despite the fact that opportunities to use "barbarous" language in user manuals are very limited, it is still worth mentioning since it is perhaps the most corrupting use of language seen today.

Orwell witnessed the corrupt, dishonest use of language by World War II propagandists as a powerful but generally underestimated weapon in the arsenals of warring governments. Controlling language gives governments the means to control thought. Orwell saw not only the Nazi's doing this, but also the British, the Americans and the Soviet.

We can only think what we have the language to think with. Without language, there is little or no conceptual thought.

Orwell saw the way governments would use terms like "collateral damage" to describe the deaths of innocent people, or their own soldiers being killed by "friendly fire" (mistakenly killed by their own side), or "ethnic cleansing" for the annihilation of whole ethic groups. This language is barbarous because it disguises the reality of the situation and makes it more likely that the public will support the action.

Barbarous terms are abstract; they do not have a down-to-earth meaning. "Collateral damage" would become horrifying if the meaning was made concrete by showing the victims as real people - perhaps one's own husband, wife or children. "Ethnic cleansing" sounds almost harmless but its real meaning is barbaric when you imagine it happening in your street, to people you know.

As an exercise, listen carefully to the way governments describe their involvement in military actions. Bear in mind Rudyard Kipling's observation that *the first casualty in war is Truth*.

Non sexist language

Care should be taken to avoid non-sexist (or non-discriminatory as it is legally known) language.

As a general guide:

- **No gender assumptions** - avoid using language which assumes a person's gender. Today, there are very few jobs where a person is always male or female. Instead of saying "he/she" or "they" when mentioning a person, refer to their job title or function, i.e. "the data entry clerk" or "the user" or simply as "you".

- **Don't get carried away** with removing apparent gender bias in language. With the best of intentions it can mutilate language. For example a "manhole" cover is the generic name of the object and to call it a "personhole" cover obscures its meaning and leaves itself open to ridicule, whereas "access" cover is acceptable.

- **Further information** - if in doubt, consult the Anti-Discrimination Act and the Equal Opportunity in Public Employment Act relevant to your state.

Writer's block

Everyone has writer's block at times, even experienced writers.

Remember those times when you were trying to write an essay or assignment? You sat and stared at a blank screen or piece of paper and the words just wouldn't come. Soon you're thinking of all the things you could be doing - some of them quite important which should probably be done right away. Next thing you know, you're doing that something else and thinking "Well I'll get back to that later". This is the gentle art of procrastination whose basis lies deep in the heart of human nature.

Preparation

The problem is often that you're expecting to hear the finished product being dictated in your mind by that mysterious process called inspiration. But before the words will start to flow you need to know a lot about the subject. So if you are experiencing writer's block, it's generally a sign that you don't yet know enough about the subject. Spend some more time preparing and getting to know the subject well.

Make a start

Another tip is to lower your expectations about the quality of output at the beginning and just write what you do know even if it sounds half baked. The important thing is to start the flow of words one way or another. Concentrate on getting as much down as possible with the intention of going back and correcting it later. It doesn't matter at this stage how bad it sounds, no one else need see it. Anything you write now can be changed later in the light of a better understanding of the subject.

An analogy is that of a log jam in a river. You simply need to get the flow going again by whatever means necessary. With writing, this is by doing a "brain dump". No dynamite is required.

Review the reference material

If that doesn't help, go back and review the reference material you have prepared. A lack of reference material as discussed in the previous chapter is often a source of writer's block. It highlights the importance of thorough presentation to the success of the documentation.

Environment

Most people work best in a quiet, comfortable environment, as free as possible from interruptions and distractions. This is easier said than done in many workplaces, particularly when the telephone never stops ringing and co-workers frequently stop by to chat.

It is important to arrange a time and a place during the working day where you can work in a quiet, interruption free environment, since you need to be able to concentrate and follow a train of thought for an extended period. This could be early in the morning or late in the afternoon when few people are about.

Routine

Writing is a mental process that can be cultivated with regular practice. Most professional writers have a rule whereby they write a certain number of pages, or write for no less than three hours every day.

The process of writing involves using the part of your mind that performs the enormously complex task of turning ideas into language. Unless you use this acquired skill regularly, it falls into disuse and does not function well. It is similar to physical fitness. Just as regular exercise keeps a person fit, writing something every day helps keep your writing faculties in good working condition.

Schedule a period each day to work on the documentation and do everything you can to stick to the schedule. If your other commitments make it difficult to allocate time on a regular basis, discuss the matter with your manager with a view to reorganising your work schedule. It is not an unreasonable request.

Ergonomics

Since writing involves sitting in one position for long periods, certain ergonomic factors need to be considered. These include the following:

Chair

Provide yourself with a chair that gives good lumbar (lower back) support. Try to avoid slouching in the chair for long periods as this places strain on the lumber vertebrae, leading in time to backache.

Screen

The screen should be on or around eye level and not closer than around 40 centimetres. Screens emit a small amount of radiation. While no definite proof exists that this radiation is harmful to humans, many people do report degrees of discomfort and eyestrain.

The intensity of radiation coming from a screen decreases rapidly the further away the screen is. Therefore, position the screen to be as close as it needs to be to allow your eyes to comfortably read the words on the screen, and no closer.

Adjust the brightness to be just bright enough rather than brighter than necessary. If the brightness needs to be high to overcome reflected light from windows, either rearrange the screen away from the direct light, or arrange blinds. All of this helps to minimise eyestrain.

Regular breaks

Occupational health guidelines recommend taking a break every hour by getting up and walking around. This not only helps your circulation and eyes, it also clears the mind.

Keyboard

Your wrists should not need to be bent while using the keyboard. Studies show that Repetitive Strain Injury (RSI) can occur where a keyboard operator, over a long period, constantly types with bent wrists. The strain is due to the tendons which pass through the wrist from the lower arm to the hands becoming inflamed because they are being stretched and constricted as they pass through the narrow aperture in the wrist known as the Carpal Tunnel.

Avoid this possibility by making sure the keyboard is not too high. Either adjust the seat higher up, or arrange for a lower desk or a keyboard drawer which fits under the desktop, or a wrist support pad.

The first draft

Before beginning the first draft, make sure the preparations discussed in the first chapter have been thoroughly made.

This chapter details both the general structure of the user guide and what information needs to be included in each part. If you follow these steps, your documentation will comply with the IEEE Standard 1063 which relates to software user documentation. Include all of them unless they specifically do not apply.

Note: Try to avoid spending too much time *editing* the first draft as it is being written. It's better to write the entire first draft from start to finish, and then edit.

Backups

It's vitally important to make regular backups of your work. Without them, it's quite possible to lose days or even weeks of work in an instant and usually for no foreseeable reason. It really does happen as most readers will know! The loss can occur for a number of reasons - a software fault, a hardware fault, a power loss at the wrong time or even the theft of the PC. And it usually happens when you least expect it.

Therefore get into the habit of saving your work every few minutes. Pressing *Ctl S* does this in most WPs. Let the pressing of Ctl S become a nervous tic.

At the end of each working day, either save the document onto a USB stick, or email it to yourself as an attachment. Do both if possible.

In my own case, I *always* email the document(s) I've been working on that day to my Gmail account. When it is in the cloud it is beyond localised disasters such as a computer being stolen or a workplace burning down, destroyed by earthquake, or inundated by floodwater.

Keeping your work in the cloud also has the advantage of being able to work on it in places other than your normal place of work provided you have access to an internet connected computer.

Some employers will not be comfortable with proprietary work products being kept in the cloud. Be guided by the applicable policies in your workplace.

Deciding on the required user documents

Prepare a documentation project plan according to the details specified in Section 2 - Preparation.

Briefly this includes the following:

- Identifying the software.
- Determining the software audience.
- Determining the document set.
- Determining the document usage mode
  - Instruction mode
  - Reference mode

User document inclusion requirements

This section outlines the information required to be shown in user manuals generally. The eleven basic components of a software user document are as follows:

1. Title page
2. Restrictions
3. Warranties
4. Table of Contents
5. List of Illustrations
6. Introduction
7. Body of Document
8. Error Conditions
9. Appendixes
10. Bibliography
11. Glossary

Table 1 below shows the inclusion requirements for specific components of a document. If a component listed as mandatory contains information not applicable to a specific document, that component may be omitted (i.e. a description of conventions may not be applicable to an overview document).

Component	Single Vol. Docs		Multi Vol. Docs	
	8 Pages or less	> 8 Pages	First Volume	Other Vols)
Title page	M	M	M	M
Restrictions	M	M	M	M
Warranties	R	R	R	R
Table of contents	O	M	M	M
List of illustrations	O	O	O	O

Introduction				
Audience description	R	M	M	R
Applicability	M	M	M	M
Purpose	R	M	M	R
Document usage	R	M	M	R
Related documents	R	R	R*	R
Conventions	M	M	M	R
Problem reporting	R	M	M	R
Body				
Instruction mode	1	1	1	1
Reference mode	1	1	1	1
Error conditions	R	R	R	R
Appendixes	O	O	O	O
Bibliography	M	M	M**	M**
Glossary	M	M	M**	M**
Index	2	2	M**	M**

Table 1: User Documentation Inclusion Requirements

M = mandatory, O = optional, R = make reference, * = should address relationship in other volumes, ** = mandatory in at least one volume of the set, with references to information in other volumes. 1 = Every document has a body, 2 = index is mandatory for documents larger than 40 pages.

User document content requirements

This section describes the required content of user documents. Specific information and the level of detail for each document are determined by the audience and the usage mode of the document. The information included in this section should be included in a user document unless otherwise noted.

The titles used in this section are general and are not intended to be prescriptive; for example, "audience description" need not necessarily be labelled as such.

Title page

The title page of user manuals should contain the following (The cover of this document is an example of the content and format of user guide title pages):

- Document title (i.e. XYZ User Guide).

- Project name (if applicable).

- Document identification number (if applicable).

- Version number (if applicable).

- Approval information.

- Optional details for the commissioning and/or accepting of the document.

- Release date - the date the version number was last changed.

- Status - whether "draft" or "approved".

- Copy no. - controlled/uncontrolled copy.

- Company/organisation logo as specified by your organisations policy on how to display logos on printed material.

Note: Header/footer is not displayed on the title page.

Inside page

The page following the title page should contain:

- Document information - including the organisational unit which produced the document, the project manager's title, name, telephone and fax numbers.

- Authors name - include mention of the people involved with the preparation of the manual. Who helped? Users need to know who to consult about the manual.

- Revision history - including at least version number, date issued, author and comments regarding the purpose/updates contained in that version.

- Include mention of the software packages used to produce the manual, i.e. Word for Windows (V6.0). This may be useful information later when upgrades or additional manuals are necessary.

- Copyright notice, for example: *Copyright © Tuffley Computer Services, 2011. This publication is copyright and contains information which is the property Tuffley Computer Services. No part of this document may be copied or stored in a retrieval system without the written permission of Tuffley Computer Services.*

Table of contents

Note: The table of contents is the skeleton of the document. It is vitally important to properly plan the organisation of the document before creating the table of contents.

A table of contents is mandatory for every document greater than eight pages in length.

For single-volume documents this requirement should be met in one of two ways.

Comprehensive table of contents for the whole document.

Simplified table of contents, with a comprehensive section table of contents preceding each section.

For multi-volume documents (i.e. a single document in multiple volumes), meet this requirement by including a simple table of contents for the entire document in the first volume. In addition, each subsequent volume should contain either of the following:

- Comprehensive table of contents for the whole volume.

- Simplified table of contents for the volume, with a comprehensive section table of contents preceding each section.

Comprehensive table of contents

Construct a comprehensive table of contents for a complete document or for a section in the following way.

- Carry entries to at least the third level of the document structure hierarchy.

- Page numbers given for each entry and are joined to the entry by a leader string (row of dots) as seen in this manual's table of contents.

Simple table of contents

In a simple table of contents include at least the first level of the document hierarchy with the corresponding page number.

List of Illustrations

User manuals may include a list of illustrations, or separate lists for different illustration types (i.e. tables, figures etc.). If included, list(s) of illustrations should appear immediately after the contents list and should contain the following:

Titles of all illustrations included in the document.

The page numbers for each entry, joined to the entry by a leader string (row of dots).

Choose whether to have separate or merged lists. The type to use depends on how easy it is to understand the resultant list(s).

Merged list - the list can be merged into one where only a few different illustration types are concerned.

Separate list - can be used to distinguish figures, tables, screens etc. where there is a significant number of each illustration type.

Headers/footers

The following information should be included in the headers and footers of user manuals:

Header contains the document ID number (refer section relating to title page), status and document version.

Footer contains, from left to right, the project name, document title and page number.

Both header and footer to use Arial 8pt normal.

Header to have a narrow ruling line below and the footer a similar line above.

Document Introduction

The following information should be given in the introduction.

Audience description.

Applicability statement.

Purpose statement.

Document usage description.

Related documents list or information.

Conventions description.

Problem reporting instructions.

The sections which follow describe these elements in detail.

Audience description

Describe the intended audience. If different sections or volumes in a set for different audiences, indicate the intended audience for each section.

In the audience description, indicate the following:

Experience level expected of the user.

Previous training expected of the user.

A description of the intended audience lets the read decide whether to read any further.

Applicability statement

Specify the following:

Software version being documented.

Hardware /operating system environment under which the software runs.

Purpose statement

Explain why the software was written and summarise the purpose of the software. Include typical intended uses of the software.

Document usage description

Describe the contents of each section, its intended use and the relationship between sections. Also provide any other directions necessary for using the document.

Example:

The manual is divided into three sections:

Section 1 Debtors Ledger used by accounts staff and management in the day to day processing of accounts receivable. Section is divided in seven sub sections which give full detail of the seven main functions.

Section 2 Creditor's Ledger used by staff in the processing of accounts payable. Section is divided in six sub sections.

Section 3 General Ledger used by staff to operate the general ledger. Section is divided in twelve sub sections.

Related documents

List related documents and give their relationship to each other. If the user documentation is made of many volumes, this "related document" information can be provided in a separate document called a "road map" or guide to document set.

Example:

Other documents relating to the XYZ Accounting Software User Guide include:

Training Guide - a set of practical hands-on exercises.

Quick Reference Guide - essential information on a laminated card.

The related documents are designed to complement not replace the user guide."

Conventions

Summarise symbols, stylistic conventions and command syntax conventions used in this manual.

Symbols - describe each symbol and how they are used in the document. (i.e. "In this manual, important points are emphasised with a heavy arrow as follows: ➔")

Style - explain any stylistic conventions with which the reader may be unfamiliar. These include conventions such as highlighting, boldface or italic script to indicate specific meaning. (i.e. "When a special term is first used, it is shown in boldface. Chapter names, menus and keyboard commands also appears in boldface when they are used in the context of performing an action.")

Command Syntax if applicable, describe command line conventions and include examples. (i.e. "Command line entries appear in Courier boldface, i.e. `XCOPY C:\SAMPLE\*.* A: /S`")

Problem solving instructions

Give the reader instructions on the following:

How to report software problems.

How to contact the help desk (if applicable).

How the reader can submit suggestions for changes in the software or the document.

List the name and contact information for the organisation responsible for responding to the problem reports or suggestions for improvement.

Body of document

Determine the content, organisation and presentation of the body of the document after determining whether the document is an instructional mode or reference mode document. In either mode, use a consistent organisational structure based on the expected use of the document, providing examples where necessary.

Body of instructional documents

Information-oriented instructional documents gives the reader background information or theory needed to understand the software. Include a scope as defined below before giving the information forming the major portion of the document. Use topics to organise an information-oriented instructional document. For example, the document could be organised as follows:

Theory.

Software features.

Software architecture.

Task oriented instructional documents gives the reader the necessary information to do a specific task or attain a specific goal. Provide all of the information listed below relating to scope, materials, preparations, cautions/warnings, method and related information. Use task relations to organise a task-oriented document or section (i.e. Organise by task groups or task sequence).

Scope. Begin this section by indicating to the user the scope of the material to be discussed.

Materials. Describe any materials the user will need to complete the task (i.e. input manuals, passwords, computers, peripherals, cabling, software drivers, interfaces and protocols). Alternatively, describe separately the materials common to all or many functions and refer to that description.

Preparations. Describe any action, technical or administrative, that should be done before starting the task (i.e. obtain system passwords, access authorisation, disk space). Alternatively, describe in a separate section the preparations common to all or many functions and refer to that section.

Cautions and Warnings. Describe general cautions and warnings that apply to the task. Place specific cautions and warnings on the same page and immediately before the action that requires the caution or warning.

Method. Describe each task, including the following:

What the user should do

What function, if any, is invoked, (including how to invoke the function and how to recognise normal termination).

Possible errors, how to avoid them and how to solve them.

What results to expect.

Related Information Provide other useful information about the task, including the following:

Tasks frequently done together and their relationship.

Other tasks customarily performed by the users of this document that could be supported by the methods of this section. Describe this support.

Notes, limitations or constraints (notes may also be placed in the specific area to which they apply).

Body of reference documents

Organise a reference mode document the way a user accesses a software function. Methods include the following:

By command

By menu selection

By system calls

Within this organisation, arrange the functions for easy access and random user access (i.e. alphabetical order or a menu-tree hierarchy). For each function, include all of the information listed below relating to purpose, materials, preparations, inputs, cautions/warnings, invocation, suspension of operations, termination of operations, output(s) error conditions and related information.

Purpose. Describe the purpose of the function.

Materials. Describe any materials the user will need to complete the task (i.e. input manuals, passwords, computers, peripherals, cabling, software drivers, interfaces and protocols). Alternatively, describe separately the materials common to all or many functions and refer to that section.

Preparations. Describe any action, technical or administrative, that should be done before starting the task (i.e. Obtain system passwords, access authorisation, disk space). Alternatively, describe in a separate section the preparations common to all or many functions and refer to that section.

Input(s). Identify and describe all data required for the correct processing of each function. Use one of the following methods.

Describe inputs used only by a single function in the section devoted to that function.

Describe in a single section or in an appendix inputs used by multiple functions. Refer to that section or appendix when describing these functions.

Cautions and Warnings. Describe general cautions and warnings that apply to the task. Place specific cautions or warnings on the same page and immediately before the action that requires the caution or warning.

Invocation. Provide all information needed to use and control the function. Describe all parameters. Include the following:

Required parameters

Optional parameters

Default options

Order and syntax

Suspend Operations. Describe how to interrupt the function during execution and how to restart it.

Termination of Operation. Describe how to recognise function termination, including abnormal terminations.

Output(s). Describe the results of executing the function, for example.

Screen display.

Effect on data or files.

Completion status values or output parameters

Outputs that trigger other actions (i.e. mechanical actions in process control application.)

Provide a complete results description for each function. If several results are possible, explain the situations that produce each.

Error Conditions. Describe common error conditions that could occur as a result of executing the function, and describe how to detect that the error has occurred. For example, list any error messages displayed by the system. (Error recovery need not be included here if they are covered in the later section "Error messages, known problems etc'; - see below)

Related Information. Provide other useful information about the task, including the following:

Tasks frequently performed together and their relationship.

Other tasks customarily performed by users of this document that could be supported by the methods of this section. Describe this support.

Notes, limitations or constraints (notes may also be placed in the specific area to which they apply).

Error messages, known problems & error recovery

Provide a list of error messages in a convenient location, i.e. in a separate section, chapter or appendix.

A complete error message index includes the following:

A list of each error message and/or code with an explanation for each.

The error that caused it.

How to recover from it.

The action(s) needed to clear it.

Describe known software problems here or in a separate document and provide alternative methods or recovery procedures.

Appendixes

Include in appendixes any material which supports or augments the information given in the main part of the manual. The list that follows is exhaustive and some points may not apply except in highly complex and/or technical situations.

Input and output data or formats which are used in common by a group of functions.

Input and output codes (i.e. inventory codes).

Interactions between tasks or functions - i.e. a description, illustrated with diagrams of the relationships and interaction between groups of tasks or functions.

Global processing limitations if applicable.

Description of data formats and file structures.

Sample files, reports or programs.

Bibliography

If applicable, provide a list of any publications that may have been quoted or referenced in the manual. Where applicable, other sources containing related information can also be included to point a curious reader in the direction of further information.

The bibliographic entry includes:

Author's name and initials.

Year of publication

Title in italics.

Publisher's name.

The second and subsequent lines are indented.

Sample format (APA):

Herzog, F. 1989. *Systems Analysis & Design*. Brisbane: University of Queensland Press

Glossary

List alphabetically in the glossary definitions of the following:

All terms, acronyms and abbreviations used in the manual with which the read may be unfamiliar.

All terms, acronyms and abbreviations with which the reader may be unfamiliar (i.e. the way in which they are used or their context).

Index

Development an index, based on key words and concepts, for any user document greater than eight pages in length.

Construct the index as follows:

Indicate the importance of information, place minor keywords under major ones. (i.e. "Printer" could be further divided into cabling, connection to, error messages and fonts).

Give the page location to the right of each entry.

List location references in one of the following ways:

By page number

By section or paragraph number

By illustration number

By other index entry

Use only one level of index entry. Where an entry points to a second index entry, the second entry should give the location and not point to a third entry.

User documentation presentation requirements

The document presentation requirements are outlined in detail in section 5 of this guide.

The following are general guidelines:

Consistency

Use terminology and typographic and stylistic conventions consistently throughout the document or set of documents. Identify any deviations the first time they appear.

Terminology

Define all terms requiring a glossary entry, acronyms and abbreviations when they appear for the first time. This means the glossary appears twice; once in a section of it's own and again distributed throughout the text as the term is used for the first time.

Referencing related material

If related material is placed in separate parts of the document, or in separate documents of a set, repetition of the information can be avoided by providing specific references to the related information.

Change control

This-version function changes

Software changes made while documentation is being written which need to be reflected in the documentation. You need a process by which changes in the functionality of the software being documented are made known to you so that they may be reflected in the documentation.

This procedure often involves the documenter receiving a copy of the software change control form which details the nature of any changes made.

Next-version function changes

Software functionality changes made while the documentation is being written but which are not to be reflected in the documentation on publication. The changes will be reflected in subsequent publications.

The distinction between "this-version" and "next-version" is usually made on the basis of a cut-off date - i.e. any changes after an agreed date will go into the next version.

Document production

This chapter outlines the recommended document production standards for user manuals. It is suggested that you use the document template supplied with this manual *(userman_template.docx)*, downloadable from:

http://www.tuffley.com/template/userman_template.docx

It is recommended that you save and rename this file as your working draft manual.

This template provides the basis for your documentation project. It conforms to the details specified in the sections below.

Document orientation

All material in the user guide with the exception of appendixes should be presented in "portrait" format. Mixed orientation in the body of the document simply confuses the user, and requires constant turning of the document to read the contents.

Copies of reports output by a system should be included in appendixes and so may be produced in "landscape" format due to their width.

Document organisation

The user manual should be organised according to the order outlined in the earlier section called "User document inclusion requirements" (Section 4).

Document format

This section outlines a uniform set of formatting standards which enable a reader to find the information they need with a minimum of time and effort.

Double sided

User guides should be produced as double sided documents from a single sided master. This reduces the bulk of the user reference guide, and makes the user reference guide easier to read as this adopts the normal "open book" format familiar to most readers.

When this format is used, it makes it a simple process to display input screens on the left-hand pages, with the relevant instructions for the input required in each field on the right-hand pages. This greatly assists the user when they are referring to the user reference guide, or using it as a training aid for new staff.

Table of contents

The document contents list is located after the preface page and should comply with the following:

Be titled "Contents" not "Table of contents".

Contain the headings from the document broken down and arranged into a logical groupings.

Contain all headings up to but not exceeding level four headings. The document itself should have no more than four levels of headings.

Illustrations lists should be shown.

Appendix(es) should be shown.

Page numbers are given for each entry and are joined to the entry by a leader string (row of dots).

TOC text should use the following font attributes:

Level	Font	Indent	Size	Space above/below
toc 1	Arial bold	1.0 cm.	12 pt.	12/6 pts.
toc 2	Arial	1.5 cm.	11 pt.	6/0 pts.
toc 3	Arial	2.0 cm.	11 pt.	3/0 pts.
toc 3	Arial	2.5 cm.	11 pt.	0/0 pts.

Level one TOC entry should have a narrow ruling line beneath.

Header/footer.

If the document being produced does not have a table of contents, it is good practice to design the document before beginning work by making a list of the headings which will form the contents list.

As a general guide, the organisation of a document should reflect a logical sequence, whether a chronological sequence, a progression from the general to the specific, or other meaningful sequence.

Page size and margins

Page set-up should have the following attributes:

Page size - A4.

Margins (left & right) - 2.54 cm.

Margins (top & bottom) - 3.17 cm.

Header & footer - 1.25 cm in from edge.

Gutter - zero cm.

Headings

Document headings should have the following attributes:

No more than four levels of heading should be used.

Headings should be indented according to their level of significance in accordance with the following table.

Lev	Style	Font & Size	Space above/below	Ruling line	Indent
1	Heading 1	Arial 18 pt bold	18/18 pts.	3 pt.	0 cm.
2	Heading 2	Arial 16 pt bold	10/12 pts.	1.5 pt.	1.25 cm.
3	Heading 3	Arial 14 pt bold	12/6 pts.	-	2.5 cm.
4	Heading 4	Arial 12 pt bold	12/3 pts	-	3.75 cm.

Level one and two headings should have a ruling line beneath, as shown in the above table. The ruling lines should be 4 pts. beneath the text and extend from the beginning of the heading across to the right margin.

All headings should have the first word capitalised and subsequent words in lower case. Proper nouns (i.e. Queensland Rail) have the first letter capitalised wherever they appear in a heading.

Note: This guide species the use of Arial and Times Roman fonts since these are excellent general purpose fonts available on most computers. As a general rule, stick to these two fonts where possible. If unavailable use another similar font. Avoid the use of unusual or fancy fonts unless you have a specific need.

Body text

The body text should have the following attributes:

Be left justified (i.e. aligned to the left, leaving a ragged right edge).

Times Roman (TT) 12 points.

Indented 3.75 cm from the left margin.

Space - zero points above and six points below.

Bullets

Bullets are used for lists where the order is not important. Bullet points should have the following attributes:

Times Roman (TT) 11 points.

Indented 3.75 cm from the left margin so that bullet character is in line with body text.

Bullet character a 10 pt. filled dot.

Hanging indent of 0.5 cm from bullet character to start of text.

Space - zero points above and three points below.

Bullet points should only be used for lists of two or more points (i.e. no single bullet point).

Representing screens

Overview

Generally, putting screens into your manual is to be encouraged, but the decision to include them will be influenced by several factors:

Expertise with computers - the reader's expertise with computers and software. Screen images are good where the reader is not comfortable or familiar with computers. Screens are helpful in showing the reader what they should be seeing. Where the reader is highly experienced, screens are less important unless;

Specific Information - there is specific information on the screen about which the reader should be aware. In this case the same information needs to be highlighted in the main text as well as the screen.

Maintainability - manuals with screens usually need more maintenance than those without. That's because those without are less likely to be affected when changes are made to the software. Think about how much time and effort is likely to be available for maintenance. Since inaccuracies like incorrect screens are a quick way for people to lose confidence in a manual, it's better to leave screens out if there's not enough time to maintain them properly.

Screen capture techniques

If you do decide to include screens, you need either:

Screen capture - a screen capture program which takes a snapshot of whatever is on the screen at a given time and saves it as a file which can then be imported into your document. There are several different products on the market which do this, ranging from inexpensive shareware up to sophisticated products that allow the file to be saved in a wide variety of formats.

Windows - if you're familiar with Windows and the software to be documented is capable of running in a window, a utility is available on the menu that appears when the button at the extreme top left corner of the window in which the application is running is pressed. Select the option to "Edit" from this menu, then "Mark Area" to mark out with the mouse cursor the area to copy. The marked area is now in the clipboard and can be pasted into Notepad where it can be saved as an ASCII file.

Screens that are saved as images are usually much larger than those saved in ASCII format. The document quickly becomes very large (i.e.. over 100Kb) when more than a few images are included in it. So saving the file in ASCII format is recommended since it takes up less space.

Screen requirements

Screens include data input screens, enquiry screens, and job submission screens.

The information displayed in screens should be meaningful but not refer to actual people or situations. Likewise, no sensitive information should be displayed.

Screens should have the following features:

Be indented so that the left edge lines up with the body text, and it should have ruling line around.

No shading - it is not necessary to shade the screen to give it more of a "screen-like" appearance.

A non-proportional font like courier should be used to represent screen text where possible.

The screen should be labelled at the bottom left corner of the box. This label should be highlighted in bold, and show the business function name followed by the screen name. For example: **Transfer Menu: Set up a Transfer - Transfer Details**

Screens should not carry over onto the next page.

Keyboard function keys

Keyboard function keys (including program function keys and keyboard control keys) should be represented in upper case bold enclosed within angle brackets. For example:

<ENTER>

<CTRL>

<PF1>

<END>

<HOME>

Numbered lists

Numbered lists are used where the order is significant, such as in procedural steps.

The attributes for a numbered list are the same as for bullet lists, with the substitution of Arabic numbers (i.e. 1, 2, 3) for bullet dots and a hanging indent of 0.75 cm after the number to accommodate double digits.

Figures & numerals

Decimal numbers less than 10 shown as numbers (i.e. 6.5). Whole numbers less than 10 are spelt in full. Spaces are used for numbers with four or more figures (i.e. 10 000).

Justification

Justification styles should be as follows:

All heading, body text and bulleted/numbered lists and captions left justified.

Table headings and table contents may be left, right or centred.

Italics

Italics should be used for the following:

To add *emphasis* to words, phrases or sentences.

Foreign words (i.e. *ad hoc*).

Titles of publications (i.e. *Concise Oxford Dictionary*)

Emphasis

Emphasis can be achieved in the following ways (in ascending order of emphasis):

Italics.

Bold Italics.

Callout character - as follows:

Note: Add emphasis to a paragraph with a callout character. Words within such a paragraph may be further emphasised with *italics* or ***bold italics***.

Hyphenation

Hyphenation is not obligatory. Many people prefer not to have body text hyphenated since it slightly increases the difficulty with which a reader scans a line, although the difficulty in many cases is negligible.

Where done, hyphenation should be made as follows:

The break is made after a vowel with the second part beginning with a consonant (i.e. peo-ple).

For multiple part words the practice of hyphenating words is decreasing, so as a general rule don't hyphenate if in doubt. For example, cooperate multi tasking.

Fractions are hyphenated (i.e. two-thirds, five-eighths). Compound words like day-to-day task-sharing are hyphenated.

Enable automatic hyphenation where this is available in your word processor.

Acronyms & jargon

While the use of acronyms and jargon can be justified in some cases, they should be avoided where possible. Where acronyms are used they should be defined in the definitions and acronyms section and also defined the first time they are used in each chapter.

Only acronyms used in the current document are to appear in the definitions & acronyms section.

Widow/orphan protection

A "widow" is the last line of a paragraph that appears alone at the top of a page. An "orphan" is the first line of a paragraph that appears alone at the bottom of a page.

Use the widow/orphan protection option in your word processor to prevent this occurring in the document.

Paragraph numbering

While headings can be numbered, avoid paragraph numbering the body text unless there is a specific need to use it.

Other considerations

Abbreviations - use fullstops with abbreviations and contractions. For example Fri. for Friday.

Punctuation Spacing - one space only after any punctuation mark.

Editing & proof-reading

This chapter outlines what to do after you've written the first draft based on the guidelines in the previous chapter.

Editing

The task of editing begins after the first draft is complete.

When formatting the first draft, use the principles outlined in the previous chapter as a guide.

Print first draft

The first draft is generally entered into a word processor and the result formatted and printed using either:

The word processor's own desktop publishing abilities.

A desktop publishing product to accept the text produced by a word processor.

Note: For some time now, there has been convergence between WP and DTP whereby all but the most demanding top-end publishing jobs can be accomplished with a word processor. For high-end jobs, Framemaker is a solid choice, though there are others.

The choice is influenced by what is the standard in your organisation. You are normally bound to follow the standard unless a compelling reason not to is given.

If you intend using a DTP product, do not format the text as it is typed into the word processor. Any formatting that is included will need to be removed or altered when it's imported into the DTP product.

Binding

Once the first draft is formatted and printed, put it in a binder. This is useful because:

Organising the material makes editing easier.

It gives you a preview of the finished product.

Edit first draft

Editing marks a turning point in your attitude towards the document. It calls for a shift from a subjective to an objective point of view. You need to shift your perspective from being the one who wrote it to the one who is now going to criticise it.

While nobody likes having their work criticised, being your own harshest critic raises the standard of your work above average and so helps to create a professional result.

Imagine you are the reader, then look for:

Clarity. Whether the manual can be understood or not. This includes whether the language is appropriate for the reader, the information is organised logically and clearly and the layout helps not hinders the reader to find information.

Brevity. Ways to be more concise. The first draft can always be made more concise, since the object has been to get the information down first, then worry about polishing it later. The reader will appreciate your efforts here. Think of how frustrating it is when you've been made to read an entire paragraph or page of convoluted double-talk when the real message could have been said in a sentence. It's a waste of everyone's time.

Active voice, present tense. As mentioned earlier, active voice present tense language has more impact and immediacy, and that's good because it helps to get the message across.

Sexist language. Check for unwarranted gender assumptions and language which may cause offence due to gender bias.

Illustrations & Examples. Are there enough of both to support and clarify the information being presented?

Editing is time consuming when done properly.

Edit second draft

After making the changes arising out of the first edit, print and bind it in the same way as above. While the first edit looks for improvements to the language, format and content, the second edit looks for technical accuracy and consistency of presentation.

Ask yourself "Have I delivered what was promised?".

Spell/grammar check, TOC, index

After implementing the changes from the second edit, do a thorough spell-check, and if you have it available, also a grammar check. Grammar checking programs allow for different kinds of writing (i.e. general, legal, business, technical etc.) Make sure you select the right category.

Also generate the table of contents and the index.

First proof

After doing the checks and generations mentioned above, print and bind the result. It is now ready to be proof-read.

Checklist

- **Acronyms.** Check that uncommon acronyms are written out in full the first time they appear in each chapter.

- **Bulleted Lists.** Check for correct indentation and consistent bullet size both within and between lists.

- **File Names, Path Names, Operator Entries.** Check for consistent use of typeface.

- **Headers and Footers.** Make sure section and chapter titles in headers and footers are correct

- **Headings.** Check for proper and consistent capitalisation. Also check for consistent phrasing of headings.

- **Index.** Check accuracy of all references by looking them up in the manual.

- **Key Names.** Check for consistent use of typeface and abbreviations (i.e., Ctrl for Control).

- **Measurements.** Make sure units of measurement are used consistently (i.e., all SI units followed by imperial units in parentheses).

- **Numbered Lists.** Check for correct numbering sequence; look for repeated, missing, or out of order numbers.

- **Page Numbers.** Make sure pages are in the correct order.

- **References**. Check for consistency of typeface and phrasing when referring to other publications and other parts of the manual. Check that references to page numbers, figures, and tables are correct.

- **Spelling**. Run your electronic spell checker AND check manually (neither method is accurate enough by itself).

- **Symbols**. Make sure symbols are used consistently (i.e., if you use a symbol such as a square to mark the end of a numbered list, make sure it appears at the end of every list).

- **Table of Contents.** Make sure page number listings match page numbers in the manual.

- **Word Choice.** Make sure wording of instructions is consistent (i.e., do not use "choose" and "select" to mean the same thing; use one or the other).

Proof-reading

Proof-reading begins after the editing is complete. From a practical point of view, the manual is as finished as you can make it without being checked by a third party.

Who proof-reads?

Finding the right people to proof-read your almost complete manual is the final challenge to overcome. Print out four copies of the manual, place them in binders and write "Draft: dd/mm/yy. Copy No. n of 4" on the cover to identify it as a draft produced on a certain date and the copy number out of the total of four.

The process happens on these four levels:

Accuracy - the information given in the manual is checked for technical accuracy. Approach the technical people who developed the software with your most polite request for help. Most of the time they aren't obliged to help and many technical people aren't thrilled by the prospect of proof-reading a manual. It's often useful to talk to the technical manager first and ask who might be able to help. Then see the nominated person and begin by saying "I was talking to (your boss) and he suggested I see you about having a look at this manual . . .". Explain that it is the technical detail you'd like them to check. Spelling and punctuation will be checked by someone else.

Spelling & Grammar - while the word processor's spell/grammar checker is a useful first pass, it's still advisable to have a literate person check the manual for spelling and grammar. In addition to general checking, ask them to look for unnecessary use of passive voice and past/future tense.

Usability - the manual is checked for usability by the same people who will use it. As with the technical proof, begin by talking to a manager in the user area and explain that for their own benefit you'd really appreciate it if someone could check the manual from the user's point of view. This can also be done as part of the "Acceptance Testing" stage of the software development.

Says too much - the manual is checked for content which might inadvertently give offence or disclose information of a sensitive nature.

Deadlines

Make it clear to the proof-readers that their help is appreciated. At the same time ask whether a month (or whatever, depending on the size of the manual and your own timeframe) will be long enough. If their reply is along the lines of :

"it should be" say you'll get back to them in three weeks or so. Always give the proof-reader a specific deadline and then diplomatically check with them a week or so before the deadline.

"I don't think so", ask how long they think it will take. If that's too long (remembering it almost always takes people longer than they say) then consider asking someone else.

Make the changes

While most suggested changes should be implemented, there will be some that don't, in your opinion, improve the manual. If this happens try to resolve it with the person concerned, otherwise ask a third party who has the authority to settle the matter. Remember, whatever the input from other people, it's you that takes the ultimate responsibility for the manual.

Reviewing & field testing

This chapter deals with the important subject of getting useful feedback about the usefulness of the document from people who are qualified to comment.

Why do we need to do these reviews and tests? The main reason is that no-one, not even the most selfless among us, are completely ego-less when evaluating their own work. Since we don't want to find fault with our work, we tend to overlook faults when we see them. In other words, documenters who see their manuals as extensions of their egos are not going to try and find all the errors in their manuals.

Reviewing the manual

As ego less documenters we need to include a review stage so that the end result will better suit the user.

Guidelines for conducting reviews:

- Select reviewers and the time when the review will take place.
- Show the reviewers ho to conduct the review.
- Give feedback to the reviewers.

Selecting reviewers and when to review

More often than not, it is necessary to find several different people who can bring a different perspective to a document.

That is unless you are lucky to have a person who can adopt multiple perspectives when reviewing a document.

The following perspectives need to be brought to bear:

Technical perspective - checking for technical accuracy and completeness.

Management perspective - checking for how well the document achieves it's overall objectives, how well it projects a positive image of the organisation and how well it protects proprietary information.

Editorial perspective - checking for conformity with accepted writing conventions.

User perspective - checking the document for it's "usability".

The matter of whether to circulate a separate copy to each reviewer or give each reviewer the same copy, each taking their turn, can be tricky. While it saves time to circulate multiple copies, you run the risk of being caught in the middle between two reviewers who disagree on the same point. Giving each reviewer the same copy, one at a time, might take longer but it reduces the risk of conflict.

Where a conflict exists, arrange for a general review meeting and ask them to resolve the issue.

Show reviewers how to review

The reviewers need to be given specific guidelines on how to conduct an effective review. Unless people have had specific training in this field, they are unlikely to know how to do it properly, since it does not feature in school or university curriculums.

As a general rule, the quality and quantity of the reviews you receive are directly related to the amount of time and effort you spend spelling out what is needed from the reviewers.

The first step is to convene and initial meeting, or circulate a initiation memorandum. At the meeting or in the memo, you should outline the general picture: what your purpose is, who the audience is, what special techniques your manual uses, etc.

One major problem which often occurs is that the reviewers confuse which perspective they should be using. The technical reviewer might slip into editorial perspective, while those looking from a management perspective can sometimes do the work of the technical reviewer.

A good way to instruct the reviewer is to place the instructions in the form of a checklist on the coversheet. The checklist could be along the lines of the following:

Is the purpose of the material clear and accurate?

Is the definition of the audience's needs and experience complete and accurate?

Do the graphics suit the audience?

Has all required information been provided?

For each review type (technical, management etc.) a separate checklist in addition to the one above should be included. For example, with the management review:

Does the information show accurately the benefits of the system?

Can all promises made in the document be kept?

Does the material protect the confidentiality of our organisation proprietary information?

As a final courtesy, when the document has been finalised, send a copy to each of the reviewers with your compliments and thanks.

Give feedback to reviewers

Regardless of how much time and effort a reviewer has spent, they should all be given feedback. Let each know the changes that will or won't be made based on their review comments. Feedback will increase the chances that the person will agree to review other documents in the future.

Field testing the manual

In field testing, users of the documentation are given the chance to use it and test it's ability to stand alone. The aim is to let real world experiences improve the quality of the finished product.

Guidelines for conducting a field test are as follows:

Select testers and the time when the test will take place.

Show the testers how to conduct the review.

Give feedback to the testers.

Selecting testers and when to test

When the manual is close to looking like the finished product. Don't wait until the document is finished, it needs to be field tested before final release.

Field testing takes a minimum of three weeks and may take six to eight weeks for complex systems

Show testers how to test

Since you won't be there to personally instruct the testers, you need to develop an explicit set of instruction for them to use.

The testers need to be prepared psychologically for their task by means of a letter that contains the kind of material discussed in the previous section on reviewing. For example you need to take the time to explain to the testers the general picture, what the purpose is, etc.

in addition, a coversheet should accompany the documentation. The coversheet should contain the following:

Identify the document as a field test document that is to be used for evaluation purposes only and not for final release.

Explain the purpose of the test.

Explain how long the test will take, and indicate the document must be returned to you at the conclusion of the test in order to avoid the document from "living on" past their useful lives.

Explain how changes and suggestions should be made - on the draft copy, in a meeting, over the phone or whatever.

Give feedback to testers

Just as it is important to make the reviewers feel appreciated, it is likewise so for the field testers.

Let each know the changes that will or won't be made based on their test results. Feedback will increase the chances that the person will agree to test other documents in the future.

Production

This section explains the "production" or printing stage. It assumes your manual is black and white since colour is both relatively rare and expensive for most user manuals.

Paper size & weight

Size User manuals should be printed on white A4 paper, preferably on both sides. This is regarded as the most practical size for presentation, ease of reading, and binding. Since it is a standard size, the masters do not need to be photo reduced before production, thus improving the manual's "readability". It can also be drilled and bound in standard A4 binders, and is a size familiar to the users.

Weight Pages should be copied or printed on 110 gsm paper. 110 gsm is relatively heavy duty and is particularly good when the ring-binding method is used, since pages can quite easily be torn out of a ring-binder by a heavy-handed or perhaps frustrated user. In general though, this is the most practical and durable weight for any printed matter that is handled frequently.

A blank buffer sheet of 150 gsm paper can be used as the first and last page of the user reference guide. This prevents the photocopied pages adhering to the plastic covers of the binders.

Print quality

The master copy of the manual should be printed on at least a 300 dpi, preferably 600 dpi laser printer. The master should be checked for degree of blackness, smudging, and paper quality. It is imperative the master is of sufficiently high quality to copy successfully.

Binding

User reference documentation should be bound in loose leaf four ring binders. These binders should be white plastic with clear plastic insert pockets on the front and spine. This means the document can be easily updated, and additional pages or chapters slotted in later at minimal cost. It allows binders to be reused as required.

(Four ring binders have been selected as the standard, as the contents are less easily torn out or accidentally damaged, than in a two or three ring binder.)

Drilling

All material included in any user reference guide, or subsequent update, should be drilled prior to distribution. All pages are to be drilled on the left-hand side in the standard position suitable for four ring binders.

These holes will be drilled no less than 1.5 centimetres from the left-hand edge of the paper.

Tab dividers

Each chapter of the user reference guide should be separated by a tab divider that clearly indicates the end of one chapter, and start of the next.

Ideally these dividers should be clearly labelled with the name of the chapter that immediately follows it, as it is listed in the master table of contents.

These tab dividers should be made of a durable material such as plastic. This ensures they do not accidentally tear out, and can be gripped to open the user reference guide at the section required.

Registration

A Project Document Register should be established as soon as the project is commenced. All project documents (including external project documents, correspondence, memos, etc, but excluding Email) should be recorded in the project document register.

Project documents from external sources should be given an appropriate document identification number to allow them to be placed in the project filing system.

Project documents should be listed in ascending order of document number, within document subject, within document type.

The Project Document Register should contain the following information for all project documents.

Document identification number.

Document title.

Document location - the full directory path and filename of the working copy of documents which exist as soft copies. The words "Paper only" (or similar) for documents which exist as a hard copy only.

Document version - the version number of the latest edition of the document.

The Project Document Register should enable staff to perform the following:

Allocate the next available document number within a given subject to a new document and record the allocation of the number to ensure each document number is unique.

Locate any project-related document.

The Project Document Register should be implemented by one of the following methods:

A hard copy document.

A software application such as a document table, spreadsheet, or database program.

Note: Where there is the likelihood that unauthorised (therefore unregistered) copies of a manual will be made, one safeguard is to inscribe the serial number using a red marker pen. Since red does not reproduce well in photocopiers, it will at least be apparent by looking at the cover of a manual whether it is an unauthorised copy.

Identification of changes

As changes are made to a user guide and a new version of that document is produced, the changes should be indicated in the revision history record. The information should be sufficiently detailed to identify each area of change.

Optionally, the changed paragraphs may be marked with revision marks. New text will be underlined and text to be deleted will be marked with strike-through characters.

Document release

Approved or unapproved documents are released to users, project team members and others on a need-to-have basis. In other words you issue as many as necessary, but no more. The more documents issued, the harder it is to control them with regard to making sure everyone gets updates.

The project manager authorises the release of project documents.

Version control & re-issue of documents

Documents which are subject to version control, and which are issued as controlled hard copy, should be issued in one of two ways.

The total document is re-issued each time a new version is produced. This is the most straight forward approach, and only requires the recipient to dispose of the earlier version and commence use of the new version. The revision history record need only deal with the version number of the total document and the changes in each version.

Only updated pages are re-issued. This requires the maintenance of version control at the page level, and the recipient should remove and replace the revised pages. It is more difficult for the user to identify if they are reading the most recent version of a page. The revision history should deal with individual pages. Where done, the additional page version control information is inserted before the table of contents.

The choice of method should be made depending on the type and size of the document, and the frequency of issue of new versions. If at all possible, the re-issue of the total document should be considered.

A means of avoiding these problems is to distribute the document as controlled soft copy only, and if the document is in hard copy form it is uncontrolled.

For controlled distribution documents, refer to the section of this Standard relating to controlled distribution.

Controlled distribution documents

In rare cases, a user guide will contain secret and/or highly sensitive information which require them to be strictly controlled. The following conditions are mandatory for controlled documents.

Where the document has "controlled" hard copies, those people/positions who are in receipt of a "controlled" copy should be known by means of a register.

Where a new version of a document is distributed, it should be accompanied by a notice to the relevant staff which identifies the reason for the change and the procedure for them to update their existing copy.

The new version and the notice need go only to the holders of "controlled" copies.

On-line documentation

On-line documentation is becoming more common due to the availability of easy to use on-line document generators and the increasing trend towards networking computers together.

Until recently, on-line documents - technically known as "hypertext" documents - could only be produced with difficulty, by a process not unlike computer programming.

Various options for on-line delivery, whether via the web or intranet means that it is now viable to convert paper documents into hypertext documents, place them on the network and make them available to all. This approach has some obvious advantages. Hypertext documents don't need to be printed or physically transported to the user, both of which saves time and money. And due to the ease with which an on-line document can be copied into a LAN directory, it's now possible to ensure that the latest version of a document is the one being used.

Another reason on-line documentation is becoming more common is that several of the biggest problems with early on-line documents have been solved. Early on-line documents used to be screen after screen of characters, often at poor resolution which made them difficult to read, and with no effective way of knowing where you were within the document as a whole. These days, the resolution of computer screens is much improved, which solves the first problem.

Conversion to on-line

The master copy of a manual should still be paper-based. From this paper-document however, it's then possible with some but not all hypertext generators to create the on-line document.

It's preferable to use the one document as both the master and the basis upon which the on-line version is made.

The guiding principle with on-line documentation is that it will not altogether replace paper-based documentation, but it does have an important part to play in the delivery of up-to-date user reference information.

Create the paper manual first

Whether you intend using on-line documentation or not, it is still necessary to first create the paper version. Having done this, you can then convert it to an on-line version using a suitable product. The on-line version should change little, if at all from the original.

The point is, the on-line document always happens second. The majority of the work you do will be in the creation of the paper-based master copy of the manual.

Maintenance

If you've followed the steps outlined in this book, your manual should be an effective piece of user documentation, and as such is worth maintaining so that users will continue to have confidence in it.

When to maintain

When to maintain a manual:

Add and/or amend details - when information has been added or changed. For example, in a ledger system, the account codes might change and new ones be added. When a significant number of such changes have accumulated, the updated pages are sent out together rather than singly.

Clarify information - when the information needs clarification in order for the reader to understand. Likewise accumulate these changes as above.

Procedure

How to maintain a manual:

Edit manual progressively - edit the manual as the changes happen. If you leave to accumulate for several months, it's often difficult to find enough time to do them.

Update Log - maintain a record of all changes made to a manual, including the date, details of the change and the reason for the change.

Print updated pages and place them in a folder or envelope.

Distribute updated pages - prepare a covering letter explaining the nature and scope of the enclosed updates and forward a copy of everything to each person on the distribution list.

Change control

There need to be a process by which changes in the functionality of the software after the documentation has been written are made known to you so that they may be incorporated in future editions of the documentation.

Two kinds of change control are recognised.

Post-publication software changes

Software changes made after the documentation has been published

Post-publication document changes

Changes to the published documentation which derive from either software changes or the rectification of software "bugs".

In both cases, the required procedure often involves the documenter receiving a copy of the software change control form which details the nature of any changes made.

Writing software project documentation

Software development projects are a documentation-intensive activity, or at least they are when done properly. This may came as news to some who imagine that projects only involve writing software.

Good supporting documentation covering the range of development activities (see list below) is essential to good project outcomes. Put another way, not doing the documentation is one of the main reasons why many software projects fail to achieve their objective of producing user-friendly, defect-free software. This happens because important information and activities are overlooked and not performed.

One reason why documentation is not done very well is because many software developers have a natural dislike of writing documents, something which goes back to their school days. The job can be made much easier through the use of templates. These can be structured to contain everything that needs to be thought of and done, plus contain enough advisory text that can later be deleted.

The templates outlined in this chapter have been based on the IEEE Software Engineering Standards (widely considered to be the best of their kind)

Software project documentation includes the following:

- Software User Requirements
- Software Requirements Specifications

- Software Test Documentation
- Software Design Descriptions
- Configuration Management Plans
- Software Project Agreements
- Software Project Plans
- Software Quality Management Plans
- Software Project Review & Audits
- Software Project Metrics
- Version & File Software Development Documentation
- Software Project Terms of Reference

Software User Requirements

Understanding the user's real requirements, rather than their requirements as perceived by the software developer, is absolutely critical to the development of successful information systems.

To achieve a high level of IS quality (user-friendliness), it is essential that the statement of user requirements be developed in the correct way. If this is done, the software that is produced will meet the user's needs, and will lead to user satisfaction. If it is not done, the software is likely to not meet the user's requirements, even if the software conforms with the specification and has few defects.

How to Capture IS User Requirements is an easy to use, step-by-step procedural guide to developing high quality, effective statement of user requirements for information systems. It contains the following topics:

- An outline of the requirements capture process
- Developing the Requirements List (RL)
- Developing the Statement of User Requirements
- Developing the Interface Control Document (ICD)
- Developing the Software Requirements Specification
- Obtaining Sign-off

How to Capture IS User Requirements relies on the proven methods outlined in IEEE Std 830 *Guide to Software Requirements Specifications* and IEEE Std P1233 *Guide for Developing System Requirements Specification*. It is written in 'plain English' and contains added features to enable project staff with average literacy skills to effectively develop a comprehensive statement of user requirements.

Software Requirements Specifications

Requirements collection is crucial to the development of successful software systems. To achieve a high level of quality, it is essential that the SRS be developed in a systematic and comprehensive way. If this is done, the system meet the user's needs, and will lead to user satisfaction. If it is not done, the software is likely to not meet the user's requirements, even if the software conforms with the specification and has few defects.

How to Write Software Requirements Specifications is an easy to use, step-by-step guide to developing high quality, effective SRSs. It prescribes both the format and content of the Software Requirements Specification (SRS).

How to Write Software Requirements Specifications is basically a 'plain English' version of the 1998 release of IEEE Std 830 *Guide to Software Requirements Specifications* [ANSI/IEEE Standard 830], but with added features to enable project staff with average literacy skills to effectively develop a SRS.

Any Software Requirements Specifications prepared in compliance with this How To guide will therefore also comply with IEEE Std 830.

The SRS has business and technical considerations added which the customer may or may not be able to provide in the original Requirements List. The SRS provides all relevant detail about the proposed system to enable a development team to commence the design/development phases.

Download the template. It is suggested that you use the handy template supplied with this book *(srs_template.docx)*, downloadable from:

http://www.tuffley.com/template/srs_template.docx

It is recommended that you save and rename this file as your working draft document.

Software Test Documentation

Systematic and comprehensive testing is known to be a major factor contributing to software quality. Adequate testing is however often not performed, leading to a higher number of software defects which impact the real and perceived quality of the software, as well as leading to time and expense being spent on rework and higher maintenance costs.

How to Write Software Test Documentation is a procedural guide to developing high quality software test documentation that is both systematic and comprehensive. It contains detailed instructions and templates on the following test documentation:

- Test Plan,

- Test Design Specification,

- Test Case,

- Test Procedure,

- Test Item Transmittal Report,

- Test Record,

- Test Log,

- Test Incident Report,

- Test Summary Report,

How to Write Software Test Documentation is derived principally from IEEE Std 829 Standard for Software Test Documentation. It contains clear instructions to enable project staff with average literacy skills to effectively develop a comprehensive set of software test documentation.

Download the template. It is suggested that you use the handy template supplied with this book *(Test Plan.docx)*, downloadable from:

http://www.tuffley.com/template/Test Plan.docx

It is recommended that you save and rename this file as your working draft document.

Software Design Descriptions

The task of developing comprehensive Software Design Descriptions (SDDs) is greatly assisted by this document.

Based on IEEE Std 1016 Recommended Practice for Software Design Descriptions, it describes the:

- Software development context in which an SDD should be created,

- Minimum requirements for SDD format and content and,

- Qualities of a good SDD.

Who is this document for? The SDD is created by the System Architect or designer and is the major deliverable from the detailed design process.

What are the Prerequisites? The prerequisite document required for an SDD varies according to the size and complexity of the software product to be developed. For large systems the prerequisite is the System Architecture Specification. In this context the SDD represents a further refinement of the design entities described in the SAS. An SDD may provide descriptions of one or more design entities. For small systems, the SDD prerequisite is a Software Requirements Specification. In this context it is the single source of design solutions to problems stated in the SRS.

Who uses the SDD? The SDD is the primary reference for code development. As such, it must contain all the information required by a programmer to write code.

Contribution to Software Quality A structured and comprehensive approach to software design is known to be a major factor contributing to Quality. Adequate design is however often not performed, contributing to a higher number of software defects which impact the real and perceived quality of the software, as well as leading to time and expense being spent on rework and higher maintenance costs.

Download the template. It is suggested that you use the handy template supplied with this book *(SDD.docx)*, downloadable from:

http://www.tuffley.com/template/SDD.docx

It is recommended that you save and rename this file as your working draft document.

Configuration Management Plans

Configuration management (CM) is the regulation of the way in which a software product evolves during the development and maintenance phases of the product lifecycle. It is the process by which the individual components of a software system are identified so that any changes to the configuration of these components can be systematically traced and therefore controlled throughout the complete span of the development cycle. CM thus forms the basis for product and project measurement.

This document is based in large part on ANSI/IEEE 1042. It describes the application of configuration management (CM) principles to the management of software development projects. CM consists of two major aspects; planning and implementation.

The objective of the CM plan is the identification and establishment of baselines; the review, approval, and control of changes to the project components; the tracking and reporting of such changes; the reviews of the evolving product; the control of documentation; and the control of the interfaces to the clients and supplier.

However it should be noted that this document specifies the minimum requirements, and therefore the project manager has the option where required, to expand and supplement as necessary for the development of specific project activities.

In specific terms, the objective of CM procedures are to provide methods for:

- Version identification

- Obtaining approval to implement a modification

- Ensuring that modifications are correctly integrated through formal change control procedures

- Controlling the identification of development status

- Ensuring that nonconforming software is identified and segregated

Template can be purchased at http://www.tuffley.com/

Software Project Agreements

This document defines the terms of the agreement between the Customer and the Developer, specifying the products and services to be provided by both parties, the specifications for products, and the delivery time and cost information.

Such an agreement is essential to the successful execution of any project, and forms the basis upon which all work is performed. The project agreement is the internal equivalent to a commercial contract.

Normally the Project Agreement makes reference to other documents which provide the detail for specific areas. When this is done the version numbers of these supporting documents is stated so that they cannot be changed without requiring a change to the Project Agreement.

This document defines the minimum content, format and presentation of a project agreement.

The Software Project Agreement document achieves the following:

- Identify the responsibilities of both the developer and the customer, during the life of the project

- Define the financial arrangements between the customer and the developer

- To document the terms of the agreement between the customer and developer so that each clearly understands what work is to be performed for what cost

- To enable a formal change control process to be applied to the project

Contribution to Quality. It has long been recognised that without a clear, documented agreement as to deliverables, cost, and schedules, the likelihood that disputes will arise between customer and developer rises sharply.

Research has also shown that where a clear and comprehensive agreement has been reached at the outset of the project, its chances of success, in terms of meeting the customer's needs, on time, within budget, also rises.

Template can be purchased at http://www.tuffley.com/

Software Project Plans

This document is a simplified version of IEEE 1058 - Standard for Software Project Management Plans.

The project plan documents the planning work necessary to conduct, track and report on the progress of a project. It contains a full description of how the work will be performed.

The benefit of using this how to guide is the consistency of presentation, enabling management to assess the plans, for their merits or limitations, more readily. In particular this how to guide specifies the format and content for a project plan by defining the minimal set of elements that shall appear in all project plans (additional sections may be appended as required).

The project plan includes the:

- scope and objectives of the project
- deliverables the project will produce

- process which shall be employed to produce those deliverables

- time frame and milestones for the production of the deliverables

- organisation and staffing which will be established

- responsibilities of those involved

- work steps to be undertaken

- budget

Using this document allows the project manager to:

- consider all relevant aspects of the project, ensuring they will be considered during the project planning stage

- produce project plans with consistent content and format

- clarify the objectives, deliverables and manner of execution of the project

Contribution to Quality. The literature of software quality widely recognises that up to 70% of IT development projects fail (in terms of either not being completed, or completed but not used by the client due to it unsuitability). One of the major contributing factors to this alarming situation is that the project was not planned comprehensively enough.

While it is not possible to foresee every misfortunes that might possibly befall a project, there are nonetheless a well-defined set of actions and attributes which if employed in the planning stage can result in all foreseeable matters being addressed. This how to guide is an easy to use checklist, as defined by IEEE 1058, and template to achieve this end. It embodies the principle of failing to plan is planning to fail..

In the same way as a systematic and comprehensive Statement of User Requirements can capture a more complete set of requirements, a project plan as provided by this how to guide allows the project manager to make sure he/she has considered all relevant matters in the planning stage, allowing them to avoid, as far as possible, unpleasant surprises later.

Template can be purchased at http://www.tuffley.com/

Software Quality Management Plans

This document is a simplified version of IEEE 730 Standard for Software Quality Assurance Plans. It specifies the format and contents of a quality plan. It identifies the practices and processes to be applied during a project to ensure that the deliverables conform to the agreed requirements.

It also identifies the quality objectives of the project, which are statements about measurable aspects of project and quality management.

The quality plan includes the:

- scope and objectives of the quality aspects of the project
- quality deliverables that the project will produce
- process by which those deliverables are produced
- organisation and staffing which will perform the quality functions

- responsibilities of those involved

The Software Quality Management Plan provides project and quality managers with a guide for the development of the quality plan. It addresses:

- quality related aspects of the project to be considered during the planning stage of the project
- the project's quality objectives, quality deliverables and how they are to be managed
- the need for consistent content and format

Contribution to Quality. As with the Project Plan, the literature of software quality recognises the importance of comprehensive planning for those aspects of a software development project that bear most closely upon its success.

Given that up to 70% of IT development projects fail (in terms of either not being completed, or completed but not used by the client due to it unsuitability), due in part to inadequate planning and execution of the project, this how to guide is an valuable aid for project planners to address the important quality-related activities. It is is an easy to use checklist, as defined by IEEE 730, and template to achieve this end.

In the same way as a systematic and comprehensive Statement of User Requirements can capture a more complete set of requirements, a project plan as provided by this how to guide allows the project manager to make sure he/she has considered all relevant quality matters in the planning stage, allowing them to avoid, as far as possible, unpleasant surprises later.

Template can be purchased at http://www.tuffley.com/

Software Project Review & Audits

This document is a simplified version of IEEE Std 1028 - Standard for Software Reviews and Audits. It encompasses the range of review and audit activities undertaken during a project.

Reviews and audits comprise the following five modules:

- Management review. The formal evaluation of a project level plan or project status relative to that plan by a designated review team.

- Technical review. The evaluation of specified software modules and documents aimed at ensuring that the software modules and documents comply with the applicable standards while conforming to the specifications. The tech-nical review may provide recommendations after the exami-nation of alternatives.

- Walkthrough. An evaluation process that can result in recommendations for improvement or identification of alternatives to the current software modules or documents being developed.

- Audit. Provide objective evidence of compliance of products and processes with standards, guidelines, specifications and procedures. Includes audits of the quality management system.

- Inspection. Rigorous formal evaluations designed to detect and identify defects in the reviewed material. Normally conducted after the event and initiated by persons outside of the project team.

The Reviews & Audits document applies to all project documents, software conduct specific reviews or audits - that need is defined by the project and quality plans. Also applies to the audit of the quality management system.

Using this document provides definitions and uniform requirements to enable project staff to perform the necessary reviews and audits of products and processes.

Contribution to Quality. An essential element in the development of project documentation is the verification and validation that their contents are true and accurate. It is recognised in the literature of software quality that the authors of documents and code inherently do not wish to find errors in their own work. They will often consciously or unconsciously overlook errors and inaccuracies. It is therefore not advisable to leave the review of project documentation to those who produce it. It is important for such checking to conducted in a consistent and systematic manner. If this is done, project documentation will serve the development process in the way in which they are intended.

Template can be purchased at http://www.tuffley.com/

Software Project Metrics

This document is a simplified version of IEEE Std 1061 Standard for a Software Metrics Methodology. It defines the standard for software quality metrics methodology. It is meant for people involved with the purchase, development, use, assistance, maintenance or review of soft-ware. The standard is especially directed at those measuring or reviewing the quality of software.

This document supplies a methodology for founding quality requirements and recognising, implementing, analysing and validating process and product software quality metrics.

This methodology applies to all software at all stages of any software life-cycle structure. Sections 1 through 4 provide, scope, definitions, and background information which is the basis of this standard; all parts of Section 5 are compulsory. Appendices A through D are incorporated for illustrative and reference reasons only.

This document does not assign specific metrics. However, the appendices include models of metrics together with a finalised example of the use of this standard.

Using this document allows the project manager to collect definitive software metrics reference for the following categories of person:

- Purchasing/project managers to identify, state and priori-tise the quality requirements for a system.

- System developers to identify definite features that should be assembled into the software in order to meet the quality requirements.

- Quality audit/assurance/control company and a system developer to review whether the quality requirements are being met.

- System maintainers to aid in change management during product development.

- User to help in distinguishing the quality requirements for a system

Contribution to Quality. Software metrics form an indispensable part of any software process improvement program. Without the means to measure improvements (or otherwise) in software quality, it cannot be determined whether process improvement is being achieved.

Template can be purchased at http://www.tuffley.com/

Version & File Software Development Documentation

This document defines these three aspects of project documentation; Document Production, Version Control and Filing.

The document production document defines the minimum requirements for the content and layout of software project documentation. A uniform 'look and feel' to project documents which is based on sound formatting guidelines leads to greater ease of use and less error and a professional image.

The version and status identification document defines the minimum information needed to maintain control over deliverables as they progress through the project lifecycle. By tagging a document and software deliverable with a version number, we are able to clearly differentiate between the current version and all previous versions. By providing the status, we are able to indicate the state of the deliverable to a more general audience. This standard therefore defines a uniform numbering and status convention applying to all versions of all deliverables. The benefit of using this standard for controlling version numbering and status is that we can achieve a high degree of consistency within printed materials and program files. A second benefit is the quick recognition of whether the material or file is an approved version or not.

The project document filing document specifies the mechanism by which project documentation is identified and stored so as to make it accessible to all those who need to have access. In particular this standard provides a document register for all project documents, and a document identification and filing system which ensures that all documents and correspondence produced or collected during the life of a project.

The document production standard applies to all project documentation. That is defined as any material generated internally during any stage of a project. It includes the following:

- Terms of Reference
- Planning Documents
- Requirement Specifications
- Design Specifications
- Business Cases
- All Quality Management System documents
- Other documents including Spreadsheets, Overhead Slides, Schedules and Presentations
- User Manuals for software products and systems
- Procedures and Work Instructions for Process Management

The version and status identification standard applies to the following:

- The numbering convention for all printed materials
- The numbering convention for all program components
- A status assignment process for printed materials and program components

Contribution to Quality. Consistent, accurate project documentation is known to be a major factor contributing to Quality. Document production, version control and filing is often not performed, contributing to a higher number of software defects which impact the real and perceived quality of the software, as well as leading to time and expense being spent on rework and higher maintenance costs.

Template can be purchased at http://www.tuffley.com/

Software Project Terms of Reference

This document defines the form and content of Terms of Reference documents. These are used to clearly define the objectives and deliverables of a given task or function, along with the parameters and constraints within which these objectives are to be achieved.

They ensure that the customer of the task or function, the organisation carrying out the work, and the individual responsible for that work clearly understand the objectives, deliverables, parameters and constraints of the task or function.

A terms of reference can apply to a wide-ranging, multi-resourced study to be carried out, right down to describing individual areas of responsibility. For example, terms of reference may be produced for a steering committee, a project, a project manager, or a task of work. In some instances, terms of reference may even be used in lieu of a project plan and quality plan.

This document defines the format and content of all Terms of Reference documents.

* Using this document clearly defines the objectives and deliverables of a given task or function and the parameters and constraints within which these objectives are to be achieved.

All Terms Of Reference documents shall be consistent in format and content.

Template can be purchased at http://www.tuffley.com/

Documentation tool support

All software requires *some* documentation, often more documentation than is commonly supplied if it is to be a successful software product. Today, there is an increasingly powerful array of tools available to help the software developer produce the required documentation.

Software developers do not often enjoy writing. There are some exceptions, people who are competent writers, and these are valuable in the IT industry because documentation is necessary and so few developers are happy to do it.

HTML tools

Hypertext Markup Language (HTML) has been the programming language of the World Wide Web since the beginning. In the past, HTML was taught as a programming language in its own right, and people could develop web pages with no more than a simple text editor.

Indeed, HTML bears a striking resemblance to the way word processing documents were marked up in the early days of personal computers. For example, in the 1980's someone using the word processor known as *WordStar* would encode a bold piece of text as *"sample text now bold now returned to normal"*. Very similar to how HTML still works.

Very few people today code HTML directly like this. They use an HTML editor, one of many available for little or no money.

HTML and related technologies

HTML editors are convenient. Some work only with HTML, others with related technologies such as CSS, XML and JavaScript or ECMAScript. Some such editors are combined with remote file transfer capabilities to manage communication with remote web servers via FTP and WebDAV, plus version control systems such as CVS or Subversion.

Types

There are several types of HTML editor: text, object and WYSIWYG (What You See Is What You Get) editors.

Text editors

HTML text editors provide syntax highlighting similar in operation to a spell checker that indicates where an error has occurred that will cause problems later.

There are also templates, toolbars and keyboard shortcuts that allow the quick insertion of common HTML elements and structures. Wizards, tooltip prompts and auto-completion may also be present.

Text editors require the user to have a certain understanding of HTML and any other web technologies the designer wishes to use (CSS, JavaScript and server-side scripting languages). This means they are not for the beginner.

Text editors such as Windows Notepad are the simplest method of all. The fle simply needs to be saved with .html or .htm extensions in order to work.

Object editors

There is a category of editor that allows for object manipulation. These are intermediate between Text and WYSIWYG.

What you see is what you get (WYSIWYG)

WYSIWYG editors produce the finished effect as seen by the user in the web browser.

This favoured method requires less technical knowledge to achieve superior results. It has not surprisingly captured the majority of the HTNL editor market. Adobe and Macromedia dominate this market.

What you see is what you mean (WYSIWYM)

WYSIWYM editors are an alternative to WYSIWYG. Instead of concentrating on format, WYSIWYM concentrates on the intended meaning of each element in the document.

Help authoring tools (HAT)

Help Authoring Tools (or HAT) are the tools of the trade for technical writers with the task of creating **online help systems.**

Help Authoring Tools take input from a variety of text editing programs, for example ASCII, HTML , OpenOffice Writer and Microsoft Word. They can also have the text entered directly into the tool.

The output from a HAT is either a compiled Help file in a format such as WinHelp (*.HLP) or Microsoft Compiled HTML Help (*.CHM), or noncompiled file formats such as Adobe PDF, XML, HTML or JavaHelp.

In addition to the input and output functions described above, HATs can also produce:

- Automatic or assisted Index generation
- Automatic Table of Contents
- Spelling checker
- Image editing
- Image hotspot editing
- Import and export of text in XML files, for exchange with computer-assisted translation programs

The most commonly used HATs include:

- Adobe RoboHelp
- Author-it
- Doc-To-Help

- FAR HTML
- Help & Manual
- Help Generator
- HelpNDoc
- HelpServer
- HelpStudio
- MadCap Flare
- Sandcastle

International communication

Communicating with people from other cultures and societies is becoming an essential skill in the 21st Century. We are increasingly coming into contact with people from a diverse set of backgrounds. It is not enough to simply say what you think and expect others to understand you. Their cultural background may prevent a clear interpretation of your meaning. Plus it is all to easy to offend people without meaning to through ignorance of their culture.

This chapter discusses two of the most commonly used models for understanding cross-cultural issues. Gert Hofstede and Fons Trompenaars are co-incidentally both from the Netherlands, but they did not work together. Their work helps us to understand how people from a particular culture *think*. If you understand this, you can communicate with them more effectively because you can adapt your communication style to suit their world-view.

Hofstede's five dimensions of culture

Gert Hofstede, a Dutch social psychologist and anthropologist, is best known for his ground-breaking work in cross cultural understanding. His most important work has been in developing the cultural dimensions theory. This theory helps us to understand (a) what is universal in human cultures, and (b) what these difference are in various specific cultures.

Hofstede discovered after administering the same questionnaire around the world that there are *five broad dimensions of culture that are present everywhere.* Think of these dimensions as having a scale from 1 to 100. Every culture in the world is positioned uniquely on each continuum. Put the five dimensions together and we have a very useful profile for understanding a unique culture, all the better to communicate effectively with them.

- Power Distance Index (PDI)

- Individualism (IDV)

- Masculinity (MAS)

- Uncertainty Avoidance Index (UAI)

- Long-Term Orientation (LTO)

Power Distance Index (PDI)

Power Distance Index (PDI) is about the degree of equality/inequality between people. To what degree do the less powerful members of a society *accept* and *expect* that power is distributed unequally?

This dimension exercises its influence largely from the bottom-up; the less powerful members endorse the arrangement and accept it as natural. Power and inequality are present in every society, and while all societies are unequal, some are more unequal than others.

For example, a high PDI society might follow a caste system in which upward mobility is very limited.

Low PDI indicates a society de-emphasises status. Equality is the collective aim and upward mobility is common.

Individualism (IDV)

The Individualism (IDV) dimension focuses on the degree to which a society values *individual* or *collective* achievement.

On the *individualist* side are societies where people have high autonomy, people are expected to take care of their own interests. On the *collectivist* side are societies where people are integrated into cohesive in-groups, usually extended families. The group cares for the individual and the individual contributes their unquestioning loyalty.

A society with a high IDV score values individuality and individual rights highly. People form relationships with others, but with the relationships is relatively weak.

Low IDV indicates a society that is more collectivist. The ties that bind individuals are strong and reciprocal, Family is very important.

Masculinity (MAS)

Masculinity (MAS) versus its opposite, femininity is about the distribution of roles between the genders. The studies found that (a) women's values differ less among societies than men's values; (b) men's values from one country to another contain a dimension from very *assertive and competitive* and maximally different from women's values on the one side, to *modest and caring* and similar to women's values on the other. In other words, it measures the degree to which societies reinforce, or do not reinforce, the traditional masculine work role model of male achievement, control, and power.

The MAS dimension considers how assertive (or masculine) versus how modest and caring (feminine) the men of a society are. The women in feminine countries have the same modest, caring values as the men. However in the masculine countries the women are more assertive and competitive, but not as much as the men. These masculine countries show a gap between men's values and women's values.

A high Masculinity score indicates that a country experiences a higher degree of gender differentiation. In such cultures, males tend to dominate a significant portion of the society and power structure.

A low Masculinity score means a society has a lower level of differentiation and inequity between genders. In these cultures, females are treated equally to males in all aspects of the society.

Uncertainty Avoidance Index (UAI)

The Uncertainty Avoidance Index (UAI) concerns the level of acceptance for uncertainty and ambiguity within a society. It considers a society's tolerance for uncertainty and ambiguity. To what extent a culture programs its members to feel either uncomfortable or comfortable in unstructured situations?

Uncertainty avoiding societies try to minimize the possibility of unstructured, novel, unknown or surprising situations by enacting laws and rules, safety and security measures at a social level. At a philosophical and religious level they do it by believing in absolute Truth.

People in uncertainty avoiding countries are also more emotional, and motivated by inner nervous energy. The opposite type, uncertainty accepting cultures, are more tolerant of opinions different from what they are used to; they try to have as few rules as possible, and on the philosophical and religious level they are relativist and allow many currents to flow side by side. People within these cultures are more resigned and contemplative, and not expected by their environment to express emotions.

Therefore a country with a high Uncertainty Avoidance score has a low tolerance towards uncertainty and ambiguity. It is usually a very rule-orientated and follows well defined and established laws, regulations and controls.

A low Uncertainty Avoidance society is less concerned about ambiguity and uncertainty and has more tolerance towards variety and experimentation. Such a society is less rule-orientated, readily accepts change and is willing to take risks.

Long-Term Orientation (LTO)

Long-Term Orientation was added after the original four to try to distinguish the difference in thinking between the East and West. From Hofstede's original IBM-sponsored studies, this difference was something that could not be deduced.

Therefore, Hofstede created a Chinese value survey which was distributed across 23 countries. From these results, and with an understanding of the influence of the teaching of Confucius on the East, long term vs. short term orientation became the fifth cultural dimension.

Long term orientation:	Short term orientation
Persistence	Personal steadiness and stability
Ordering relationships by status and observing this order	Protecting your 'face'
Thrift	Respect or tradition
Having a sense of shame	Reciprocation of greetings, favours, and gifts

Trompenaar's seven dimensions of culture

Fons Trompenaars is a Dutch researcher in cross-cultural communication. He and colleague Hampden-Turner developed a model of culture with seven dimensions that is helpful in understanding people's mind-set when communicating with them.

Universalism vs. particularism

Universalism vs. particularism describes how people judge other people's behavior.

The Universalist places great emphasis on following the rules under all circumstances. Exceptions are not made on the basis of who is involved, or particular circumstances. An example is when a policeman insists of giving a motorist an infringement notice "because it's the Law".

The Particularist decides what to do based on who is involved (is it a close friend, or a stranger) and the specific circumstances that apply in that instance.

Universalist countries include the USA, Australia, Germany, Switzerland, Sweden, United Kingdom, Netherlands, Czechoslovakia, Italy, Belgium.

Particularist countries include Brazil, France, Japan, Singapore, Argentina, Mexico and Thailand.

Individualism vs. collectivism

Ssimilar to Hofstede's dimension, Trompenaars describes an Individualistic society as a modernizing one, while a Communitarian society is oriented towards the past, respecting tradition and orthodoxy more than the Individualistic society.

These two orientations complement each other, rather than oppose. See Hofstede's dimension above for more detail.

Neutral vs. emotional

Do we display our emotions? Neutral cultures believe that it is unseemly to express emotion publicly. It is not that they do not feel emotion, they simply do not show it. Rationalism is emphasised.

Emotional cultures believe that it is fitting and right to express emptions publicly. They do this with open displays of the full range of emptions. Weddings are likely to be more joyous, funerals more mournful, and the range of emotions in between displayed with laughing, smiling, grimacing, scowling and gesturing.

Specific vs. diffuse

How separate do we keep our private and working lives? Also referred to as the *concern-/commitment-dimension*. It is expressed at the individual or personal level when a person is affected by a certain situation or action.

In specific-oriented cultures clear distinctions are drawn between the various aspects of life. Work and family and personal are clearly partitioned off from each other and do not overlap. A personal may be open in the public space but closed in the private space. They are likely to be direct, to the point and purposeful in their dealings with people.

In diffuse-oriented cultures there is no clear distinction between the aspects of a person's life. They tend to overlap and overflow into each other.

Achievement vs. ascription

Do we have to prove ourselves to receive status or is it given to us? Does a person's status in society derive from their achievements, or from their social position, family connection?

In achievement-oriented cultures titles are earned. Respect for others is based on their perceived competence. Decision-making is challenged on technical and functional grounds.

In ascription-oriented cultures a titles are inherited or ascribed to a person as a way of identifying their position in the social hierarchy. Respect for people is based on seniority and hierarchy. Decision-making is challenged by people with higher authority.

Sequential vs. synchronic

Do we do things one at a time or several things at once? This dimension considers how people in different cultures perceive time; sequential and synchronic.

Perceiving time sequentially means seeing the passage of time as a series or sequence of events. A sequential person has crucial path worked out in advance with times for the completion of each stage. People with this kind of understanding of time dislike having their plans disrupted by unforeseen events. They schedule very tightly, with thin divisions between time slots. It is impolite to be few minutes late because the whole day's schedule is affected. Time is a valuable commodity to be used wisely.

Perceiving time synchronically means seeing the past, present and future as an interconnected whole. The future and the past reflect upon and influence the present. People with this orientation tend to perform several activities concurrently, in parallel. They may have a desired goal, but no fixed idea of a sequence of events leading to that goal. Any one of several different paths will get them there.

Synchronic people may be offended by sequential types who insist on finishing one task before starting the next. After all, they, the synchronic, would have no problem interrupting their current activity to accommodate the other.

Internal vs. external control

Do we control our environment or are we controlled by it? Internal control cultures identify with mechanisms. Society is thought of as a machine that mechanistically obeys the will of its constituent members.

External control cultures see society as a product of Nature, its function and form being influenced by organic factors.

High performance teams

The days of software developers working alone have all but gone. Almost any project today is worked on by a team of developers. Sometimes the team is complex, involving multi-disciplinary elements.

Knowing how to work effectively with others to achieve a desired outcome does not come naturally or easily to many IT developers. This chapter outlines the principles of how to turn a group of people into a high performance team.

Introduction

There are many definitions of teams. This is hardly surprising given the importance of team-work stretching back to our evolutionary past.

A survey of the commonly understood meaning of 'team' suggests that it is a *co-operative unit* of some kind (humans or animal) linked by a *common purpose*. Complex projects can be comprised of multiple teams with an interdisciplinary focus.

For an empirical definition, one that is not based on the 'wisdom of the crowd', we turn to Watts Humphrey, one of the founding fathers of the Software Engineering discipline. In terms of teams engaged in the development of technology, the definition selected by Humphrey as an operational definition in this well-respected model of team software process is:

A team consists of:

1.   At least two people, who
2.   Are working towards a common goal/objective/mission, where
3.   Each person has been assigned specific roles or functions to perform, and where
4.   Completion of the mission requires some form of dependency among group members.

Trust

The concept of trust is an indispensable aspect of successful projects. Trust can be defined as *confidence*. In terms of project groups we can make a distinction between bilateral trust between individual group members (one-to-one trust) and general trust (one-to-all) in the project group.

The importance of trust to successful project outcomes is indicated by it being mentioned frequently in much of the literature on network theory.

Group interdependence

Group *interdependence* can be illustrated in what have been recognized as truly great teams. One such team was the Manhattan Project (developing the U.S. Atomic Bomb in World War II). Members of great teams submerge their egos for the mission' in the spirit of self-sacrifice. The Manhattan project team members described their leader J. Robert Oppenheimer as charismatic, a man with an 'intense presence' and 'poetic vision'. Oppenheimer was apparently capable of motivating, even inspiring the team members with a powerful team spirit. "Inspiration' (or to infuse with spirit) is used in the same sense that 'charisma' is said to be a 'gift from God' in the original Greek. Evidence of this is heard in the description by one scientist who went so far as to comment that in Oppenheimer's presence *'I became more intelligent, more vocal, more intense, more prescient, more poetic....'.* Oppenheimer as team leader/project manager acted as a kind of 'spiritual midwife' with his ability to bring forth and realise the potential of the team members.

It is not clear whether Oppenheimer acted consciously in this regard, or that he was exercising an instinct. Most likely it was combination of the two. Oppenheimer managed to create and instil among a collection of gifted scientists a sense of common purpose, interdependency and of being on an important mission together. This appeared to go well beyond what might be expected under the serious circumstances in which they found themselves. It was strongly suspected by the Americans that Nazi Germany was pursuing similar research, and not only the outcome of the Second World War, but conceivably the future of humankind depended on their being first to realise their objectives.

Team spirit and interdependence on the Manhattan Project was engendered not only through the scientific work, but also through recreational pursuits like skiing in the mountains of New Mexico, square dancing and general partying in which people had 'enormous amounts of fun'. It is not reported how readily the European members of the team adapted to American square dancing.

Virtual Teams

A virtual team can be broadly described as a group of people who perform their work using information and communication technology to bridge time, space, and organizational boundaries. In common with co-located teams, virtual teams have complementary skill-sets, are interdependent and share a common purpose. A working definition based on the relevant literature is given in the following section.

Virtual teams can include integrated teams by definition. The previous sections discuss at length the characteristics of integrated teams, and while this discussion is located in the 'co-located' team category, it should not be perceived as belonging solely in that category.

A virtual team consists of at least two mutually interdependent people, who are geographically dispersed, and who are working towards a common goal/objective/mission, where each person is assigned specific roles or functions to perform, and where communication is facilitated by a combination of telecommunications and information technologies to work towards the completion of the project/mission.

Characteristics of high performance teams

High performance project teams tend to have certain characteristics in common, regardless of who they are and the field they work in. Six high performance project teams (called 'Great Groups') were examined. These teams have one thing in common, in one way or another changed the world. These include teams at Xerox's PARC labs pioneers in human-computer interface research), the Manhattan Project (development of A-bomb for US DoD), the 1992 Clinton campaign (defeated Gorge Bush Snr fr President), Lockheed Skunk Works (innovative aircraft, including U-2, SR-71, F-117, F-22), and Disney animation studios (standard-setting animation techniques). Their examination aims to encapsulate the characteristics of successful team-work. Their findings can be summed up in the quote *'All great teams--and all great organizations--are built around a shared dream or motivating purpose.'*

Their findings into ten principles, as summarised below:

1.   **At the heart of Great Groups is a shared dream**. Great Groups are convinced they are engaged in important work, sometimes nothing short of being on a 'mission from God'. The work becomes an abiding obsession, a quest that goes well beyond mere employment. This intensely shared vision and sense of purpose endows cohesion and persistence.

2.   **Conflict is constructively managed by submerging individual egos for the greater good**. Inevitable conflict between team members is managed by reminding members of the over-arching importance of the mission. For example, during the Manhattan Project, George Kistiakowsky, a pivotal member of the team threatened to leave over conflict with another member. Project leader Robert Oppenheimer said, 'George, how can you leave this project? The free

world hangs in the balance.' Kistiakowsky was thus persuaded to submerge his individual concerns for the greater good of the project.

3. **Team members are insulated from bureaucracy.** Great Groups have a dislike and disdain for their corporate overseers, who are perceived as functionaries with little real vision who must not be allowed to interfere with the real work. The important task of insulating the team is performed by a leader – though not necessarily the same person who promulgates the galvanising dream. For example, on the Manhattan Project it was General Leslie Grove who insulated the team from the Pentagon, while Oppenheimer kept the team focused on its mission. At Lockheed, Kelly Johnson had himself appointed to the board of directors in order to insulate his beloved Skunk Works. Physical distance from corporate headquarters is an important aspect.

4. **There is a real or perceived enemy.** The presence of an enemy who must be defeated is a powerful motivator and unifier of effort. This is seen most clearly on the Manhattan Project where there was indeed the Imperial Japanese and the Nazis. Indeed, most organizations have an explicit and/or implicit objective to overwhelm their competitors. For example, Apple Computer's implicit mission was, 'Bury IBM'.

5. **Great Groups regard themselves as having underdog status.** High performance groups seem to be often constituted by unconventional people.

So-called mavericks, people at the edges of their disciplines. They are at the edge through a process of self-marginalisation, being unwilling to subscribe to mainstream values, and marginalisation by the larger discipline through their aforementioned refusal to be conventional. For these people, there is no sacred cow that is beyond criticism. 'Membership in a Great Group isn't a day job; it is a night and day job.'

6. **Team members often pay a personal price.** Membership in high performance teams, or Great Groups seems to exact a high price. Because the work is not only a day job; but a night *and* day job, it is not uncommon for divorces, affairs, and other extreme emotional matters to be common, particularly when the project finishes. At Lockheed's Skunk Works, for example, team members were not allowed to tell their families what they were working on. The energy and intensity that such groups generate can lead to unfortunate, sometimes socially dysfunctional consequences in their lives outside of the project.

7. **Great Groups produce strong leaders.** While group leaders tend to be non-hierarchical, open, and egalitarian, they nonetheless display strength. There cannot be a Great Group without a great leader -- and vice versa. In an important way, Bennis and Beiderman (1997) observe that it is the groups that make the leaders great. The willing and unified support of the group endows the leader with strength and authority. Leaders are

connoisseurs of talent, more like curators than creators.

8. **Great Groups are the product of meticulous recruiting.** Selecting the right people for a group means being clear about what is required, and how to recognize it in potential team members. Potential members can be put through a very rigorous selection process, in order that the right person goes into the right job.

9. **Great Groups are usually young.** Youth provides the physical stamina demanded by Great Groups. The average age of the physicists at Los Alamos was around 25. Great Groups are also young in their spirit, ethos, and culture. Importantly however, because they are young and naive, group members don't know what's supposed to be impossible, which gives them the ability to do the impossible. The inflexibility of mind, the setting-in of habitual ways in the minds of older people is the real enemy here, not the nominal age of a person.

10. **Successful groups deliver the goods.** Great Groups must, in the end, produce a tangible outcome external to themselves. Most groups dissolve after the product is delivered. Steve Jobs was fond of reminding his Apple team that their work meant nothing unless they brought a great product to market.

Characteristics of low performance teams

As a final word on the subject of what characterises a successful team, it is instructive to review Ed Yourdon's oft-cited 'death-march' project characteristics.

By knowing what a successful project is not, we are in a better position to understand what a successful project is.

A 'death march' is defined by Yourdon as being one in which the project parameters are exceeded by at least 50% . Said parameters include the schedule, staffing, budget and system functionality aspects of the project. While individual death march projects can technically succeed from an outcomes perspective, it is the death march culture that ultimately fails.

Yourdon cites the following reasons for death march projects:

- Politics, politics, politics
- Naïve and/or devious promises made by marketing, senior executive, inexperienced project managers, etc.
- Naïve optimism of youth: 'We can do it over the weekend'
- The 'startup' mentality of fledgling entrepreneurial companies
- The 'Marine Corps' mentality: Real programmers don't need sleep
- Intense competition caused by globalization of markets

- Intense competition caused by the appearance of new technologies

- Intense pressure caused by unexpected government regulations

- Unexpected and/or unplanned crises eg., your hardware/software vendor just went bankrupt, or your three best programmers just died of bubonic plague

'Death march' projects, or failed projects in general terms may be characterised as combining some or all of the above elements. The onus is on the project leader to take action to avoid or mitigate the effects of these elements, where present.

Improving project team performance

The following list summarises how team performance can be improved. It covers the full range of activities on teams; it should be noticed that **effective communication** lies at the heart of all of these.

1. Project managers need to understand the barriers to team development and create a work environment that works constructively with the team's motivational needs. The following barriers are particularly important: (a) bored, demotivated, disinterested team members, (b) detached management, (c) unclear goals and priorities, (d) funding uncertainty, (e) role conflict, politics and power struggle (f) poor team leadership, (g) no project charter, (h) not enough planning and project

definition, (i) poor communication between members and (j) negative conflict between members, as opposed to creative conflict that derives solutions to problems being argued about.

2. Team structure and procedures must be defined during the start-up phase.

3. Everyone understands the the goals of the project, and that they "buy into" those goals. Clear and frequent communication with senior management and the client is critically important. Status review meetings are used for feedback.

4. Carefully selected leadership for the project is essential from the very beginning. Senior management are responsible for this. The project leader must be a high credibility individual. Project staff and other stakeholders must see them as knowing what they are doing.

5. Project assignments are carefully matched to the expertise of team members, and agreement reached between all parties that the assignment is accepted. Project managers are responsible for staffing their own projects, according to their best judgment. If dual-reporting relationships are involved, staffing should be conducted jointly between the two managers. Each team member discusses the following with their manager; (a) the specific tasks to be performed, (b) the outcome, (c) timing, (d) responsibilities, (e) reporting relations, (f) potential rewards, and (g) the importance of the project to the company. Task assignments should only be made if

the person's skills are a match for the position requirements and the person shows an interest in the project.

6. The project manager takes great care that the requirements for the project deliverables are clearly understood. This means involving all stakeholders and taking sufficient time to do the job properly. This involvement leads to an (a) improved understanding of the task requirements, (b) stimulate interest, (c) help to unify the interest, (d) help to unify the team, and (e) lead to commitment to the project plan, regarding technical performance, timing, and budgets.

7. The project manager, either directly or delegated to team leaders, facilitate communications among team members, to and from senior management, and the customer/sponsor community. This is done by scheduled project review meetings, management briefings, as well as project planning and tracking activities.

8. Team building sessions are performed throughout the project lifecycle, particularly during the formation stage. The team is being brought together in a relaxed atmosphere to discuss such questions as:

   ▪ How are we going as a team? What is good, what can be improved? What do we need to do to improve?

   ▪ What are the likely future problems? Can they be avoided by taking appropriate action now? How can we future-proof the team?

- What improvements need to be made in the team environment? Do we need to modify our objectives and/or our operating procedures? Are we monitoring progress well enough?

9. Look for lack of team member commitment as early as possible in the project. Do what can be done to change possible negative views. Insecurity is often the source of negative attitudes. Try to understand why the team member feels this way, then try to reduce their fear. Conflict with other team members may also be a reason for lack of commitment. The project manager should intervene as early as possible. If a team member's professional interests lie elsewhere, the project manager examines ways of satisfying at least part of the team member's interests or consider replacement.

10. Continuing senior management commitment to the project is essential. The project leader is primarily responsible for this. It ensures that effective relations continue between the various interface groups, plus continued resource allocation to the project.

11. Performance of team members should be monitored. If problems are noticed, discussions with the team member(s) should be held as soon as practical with a view to restoring peak performance.

12. During the course of the project, the nature of the problems encountered by the team are likely to change. As old problems are solved, new problems emerge. Potential problems should be identified before they occur, and effort put into preventing their

occurrence. The amount of effort needed to avoid problems is usually much less than that which is required once the problems

13. The project manager should show *leadership*, that is, finding a way to get people to *want* to do what it is the leader wants them to do. They do this by having an exciting vision for future outcomes, showing individualised consideration for each member of the team so no-one feels like a mere unit of production.

The Ethical Technologist

If technology is powerful and pervasive, how can we make sure it remains friendly? After all, it is here to stay. We could not get rid of it, even if we wanted to. We need ethical technologists who understand the true nature of technology, beyond the clichéd images and whiz-bangery.

Toolmakers, par excellence

Going back millions of years, humans have excelled as tool-makers. Our information technology is perhaps our finest tool-set, though by no means our only tool-set. More than just inanimate objects though, information technology is an extension of our mind. Technology lets us extend our ability to think and process information beyond our biological brain, out into the environment.

This ability to extend our minds into our tools did not begin with information technology. We have always done this. Andy Clark, a respected cognitive scientist reports that brain scans show that if you were to pick up, for example, a garden rake and start to use it to gather leaves, within a short time your brain would have mapped the tines of the rake to be extensions of your hands. We call it *haptic touch*.

The computers that we have come to depend on are just another tool that we project our minds into and use to outsource some of our thinking tasks. If you doubt this, imagine if you lost your personal computer. In some ways, it would be like having a stroke. Part of your brain would have disappeared, and you would very much feel the lack of it. You might feel lost and debilitated until a replacement is found, complete with restored data.

So people have a closer relationship to information technology than is commonly realised, having become an extension of our biological mind. As millions of extended minds have reached out and merged with each other we can observe a remarkable phenomenon, the formation of a new layer of consciousness in the world.

Welcome to the Noosphere

The French philosopher Pierre Teilhard de Chardin foreshadowed the Internet as far back as the 1930's. In his book *The Phenomenon of Man* published after his death in 1955, Teilhard describes humankind as the evolutionary process of life on Earth becoming conscious of itself. The generation and exchange of ideas between people over time created a collective memory that enhanced human consciousness to the point where a thinking layer was created that enveloped the earth. He called this the noosphere after the Greek for mind. The noosphere is an extension of the biosphere, which is itself founded upon the geosphere.

As Kevin Kelly, Founding Editor of Wired Magazine pointed out in 2007, the Internet is a neural network that approximates the human brain in complexity. I'm about to quote some very big numbers. The Internet has around 55 thousand billion links, about same number of synapses in the brain. There is about the same number of transistors as neurones (one quintillion, that's a one with 18 zeroes behind it). They both have about the same amount of data storage capacity (255 billion gigabytes) and everything is firing at roughly the same frequency as each other. We can make a valid if rough comparison between one human brain and the Internet of 2007.

But the human brain is not doubling in capacity every two years. The dimensions of the Internet cited here will have doubled for every two years from 2007. Kelly predicts that by 2040, the total processing power of the Internet will exceed that of six billion human brains. One might say that Teilhard de Chardin's noosphere is alive and well and growing at a phenomenal rate. The noosphere is one explanation for how it is that ideas and inventions seem to occur spontaneously to people in different parts of the world at the same time but with no contact with each other. The history of innovation is rich with such examples.

One cannot mention this tantalising future without trying to characterise it. What will the Internet of the future be like? It will be smarter than it is now. Artificial intelligence is being built into it by major movers and shakers like Google. Each person who uses the web will be known by it, at least to some extent. The person's likes and dislikes will be known and catered for by so-called intelligent agents. These agents are semi-autonomous software programs that can move about the Internet doing things for you. An electronic personal assistant. These agents might be embodied into something pleasing to our eye and made to be comfortable to be with.

The Internet of the future will be pervasive, some might say ubiquitous. Everyday objects in our world will have webness built into them in the form of computer chips and these will be in communication with the Internet.

We are likely to become more and dependent on the Internet too. After all, as Andy Clark points out, our technology is an extension of ourselves, of our biological brain. We instinctively outsource thinking tasks into these external thinking devices and the Internet will have become one enormous thinking device.

Orwell's 1984

This vision of the future may be alarming to some people. It sounds too much like the Big Brother of George Orwell's 1984. That nightmare did not eventuate, thanks partly to Orwell's brilliant articulation of the dangers. Orwell was writing about a world traumatised by two catastrophic wars, where ideologies struggled for hegemony. Capitalism and Communism united to defeat Fascism, then went against each other once the Nazis were disposed of. Orwell was describing an extrapolated world of political surveillance 40 years into the future from the time he was writing about. Shades of the East German Stasi (**Sta**at**ssi**cherheit). It was a warning well worth heeding, as discussed a little later.

Instead of being an instrument of suppression and paranoia, the Internet is the greatest force for the democratisation of information since the printing press. It puts the accumulated knowledge of thousands of years of cultural evolution at the disposal of billions of people world-wide.

In the world of the early 21st century, Capitalism has triumphed. With the exception of some hard-liners, the world is sold on Capitalism. Peace and prosperity, consumer goods for all. A comfortable life. A good standard of living. Former foes Germany and Japan now major economic powers thanks to America's efforts to rebuild their economies post WW2. The Soviet Union disbanded, each former constituent state pursuing their own prosperity agendas. Still a couple of hard-line Communist states left standing, but for how long? China surging ahead as an economic powerhouse. Now all we have to do is figure out how to bring the developing world to the table for a share.

The Internet as it is today

Technology in the early 21st century allows people to project their minds anywhere in the world, unbounded by physical limitation. In a sense the human mind has come to encompass the entire planet. This is a fair description of the Internet, the most powerful and pervasive tool ever constructed by us humans, the cleverest of all tool-makers.

We speak of the Internet as a single entity, but it is really a vast heterogeneous entity that exists beyond anyone's control. As much as nations and global conglomerates would like to exercise control over the Internet, it is ultimately beyond anyone's control. It was designed that way. Originally conceived in the late 1960's by the US Military's Advanced Research Projects Agency (DARPA) as a fault-tolerant communications system, it had to withstand having elements of it be destroyed or made unserviceable, while the rest continued to function. At the time, telephones were the only 'theatre-scale' communications systems, but the centralised switching mechanisms were vulnerable to attack, making the whole system unreliable. There would be an obvious advantage in having digital communications that side-stepped this vulnerability in the event that the cold war became hot.

Today you can use the Internet to quickly access information on any topic, however obscure. Beyond satisfying your curiosity on any subject, you can look up someone's phone number, buy from an array of millions of items for sale, access government services, find books in your local library, read the newspaper, check the weather, the sports scores, or the price of real estate. You can do your banking, book a flight to the other side of the world, reserve a hotel room for when you get there, a rental car to drive around in, concert or movie tickets and restaurant reservations for after the show. Online maps and high resolution satellite images let you visit places and navigate through them. You can visit these places in cyberspace and make arrangements before travelling to them physically.

You can make friends on social networking sites, find romance, arrange casual sex, display your artwork and seek out like-minded people to communicate with about the most specialised interests. You can download movies and music, much to the annoyance of the recording industry and movie studios. You can spend more and more of your time in cyberspace playing immersive online games like Second Life and World of Warcraft. And this is where the addictive nature of the web is most clearly seen. Some people, particularly those with an introspective nature, are more comfortable in cyberspace and prefer to dwell there most of the time.

Susan Greenfield in her recent book *ID: The Quest for Identity in the 21st Century* speculates that people will become so accustomed to the safe, sanitised world on-line that meeting people face-to-face in the real world will be disgusting in the same way that a meat-eater today would be disgusted if they had to actually slaughter and butcher an animal to obtain the meat rather than find it in sterile cling-wrap from an air conditioned, muzak-enhanced supermarket.

What are the risks of technology?

With greater transparency and access to information globally comes the potential for abuse. But with the many benefits of information technology, it is arguable that the potential for *abuse* should not in itself prohibit the *use* of technology, as the Luddites would suggest.

We need for the technologists who are the creators of our technological future to have an mindset that moderates the potential for harm. This offers a viable way forward into a technological future.

We need to know what we are dealing with. What are the various ways that technology can be used to harm people? There is intellectual property theft where all manner of material from photos, to text, to music and videos is copied and distributed without the owner's permission. Otherwise known as 'piracy', the annual loss of revenue is reckoned by the recording industry to be in the hundreds of billions of dollars, though these estimates are considered by some to be grossly inflated.

There is pornography, a very broad spectrum of material ranging from non-violent erotica between consenting adults, to fetish erotica between consenting adults all the way to non-consensual material and the most depraved of all involving violence towards children.

Identity theft is widespread where someone's personal information is mis-appropriated and fraudulently usually to the detriment of the owner.

Privacy is a major problem where your email address and other personal information is distributed to unknown third parties without your consent. Spam is a prime example.

Cultural differences around the world mean that material that is considered acceptable in one culture is able to reach other cultures in other parts of the world, causing offence. There are also issues like stalking *and cyber-bullying, gambling, and social inequity* that gives greater access to information to some and not others.

The ethical technologist

How can we instil an ethical do no harm mindset in technologists? For the past six years I have been trying to do that with IT students at Griffith University. Such a process must go beyond the teaching of professional codes of conduct, though this is a good place to start. It must reach into every area of a proto-technologist's life, drawing the different compartments of their lives into an integrated self-aware whole. They must appreciate the human consequences of their actions, even when they occur at a distance, beyond their senses and the cubicle-constrained world in which they work.

We talk about moral philosophy and the various theories that have grown up over the past two and a half thousand years since the Classical Greek philosophers first turned their minds to the task. Some students find this interesting, but for many it is a struggle, not only to appreciate the nuanced thought of the great philosophers, but also to see the relevance of it to technology students such as they. Students who seek the comfort of knowing there is a definite right and wrong answer, are uncomfortable with shades of grey.

We review the common denominators of moral behaviour in philosophical and religious thought, those persistent recurring truths that find expression across time and cultures. We discuss how to become aware of the cause and effect linkages that permeate our lives. How one's decisions in this moment determines what happens to us in the future. Accepting that one is ultimately responsible for what happens to us as a necessary step towards self-mastery and a constructive future.

This is a theme well-explored in the literary world as well. American writer John Steinbeck said this in Chapter 34 of his 1952 novel *East of Eden*:

I believe that there is one story in the world, and only one, that has frightened and inspired us, so that we live in a Pearl White serial of continuing thought and wonder. Humans are caught in their lives, in their thoughts, in their hungers and ambitions, in their avarice and cruelty, and in their kindness and generosity too - - in a net of good and evil. I think this is the only story we have and that it occurs on all levels of feeling and intelligence. Virtue and vice were warp and woof of our first consciousness, and they will be the fabric of our last, and this despite changes we might impose on field and river and mountain, on economy and manners. There is no other story. A man, after he has brushed off the dust and chips of his life, will have left only the hard, clean questions: Was it good or was it evil? Have I done well -- or ill?

Steinbeck seems to be saying that all human endeavours, all of our thoughts and actions, can be distilled down to a single theme, the on-going struggle within us between good and evil. Our challenge is to develop enough insight into the web of causality that we are able to consciously choose the course of action that will involve good consequences. And in so doing gain more effective control over our life and where you want it to go.

How does this apply to technologists? I propose the following two simple principles for ethical IT practice. It is simply expressed because I follow Einstein's advice that the greater the truth, the more simply it can be expressed. Complex theories are beloved by academics but are lost on most students. It is simplicity that gets the message across.

A technologist's action can be said to be ethical if the person(s) affected:

- *Gives their informed consent, and*

- *Is not de-humanised in the process*

There has been extended debate in tutorials on whether informed consent alone is required. If a person does not mind being de-humanised or harmed then that is their choice and this should be respected. Arguably though, it is better that we do nothing to de-humanise others since to do so will likely result in our own dehumanisation and so impair our ability to act ethically in the future.

In other words, technologists (and people generally) need to have enough awareness of the web of causality to let them consciously choose the course of action that will involve good consequences.

Humanising technology

So the two important aspects of ethical technology development are that people are made aware of the consequences of use and give their informed consent before using it, and that the technology does not de-humanise the person in the process. Ideally, technology should enhance a person's humanity, but at the very least it should not diminish it.

The humanising influence is a more complex idea. What do we mean when we say technology must not de-humanise those that use it? It is remembering that the technology is a means to an end, and not an end in itself. People before technology. Technology the servant, not the master of people.

It is not uncommon for technologists to fall in love with the technology they create, and overlook that this beautiful artefact is not an end in itself, a thing of beauty in which they have invested themselves. Many technologists I have known regard themselves as part technician, part artist. If this tendency to regard technology as an end in itself is not curbed, it will produce a world in which people increasingly serve the needs of technology.

A humanist perspective in technology development therefore keeps the technology user-friendly, life-affirming.

The idea that technology can be a de-humanising influence in the world is not a new one. The German philosopher Martin Heidegger in his influential 1954 essay The Question Concerning Technology suggested that technology is an expression of the human tendency to exploit and mechanise the natural world. Over time, we are de-humanised in a world where efficiency and exploitation rule, and an appreciation of Nature for its own sake is diminished.

From an evolutionary psychology point of view, we humans have throughout our long past made tools that improve our chances of survival in a hostile environment. We made tools that change and control our natural environment. After hundreds of thousands of years, we became so good at it that the tools became an end in themselves. The part of the human mind that had become so good at dominance and control grew too influential, at the expense of a person's gentler nature.

So technology should not be an end itself. It is there to help people to live their lives more fully, to achieve their human potential. As Kelly points out, technology at its best can help people express their true selves and highest potential. Imagine Mozart in a world before the technology of the piano had been invented, or Van Gogh in a world before inexpensive oil-paints had been invented, or Hitchcock before the technologies of film. Today, there are millions of children being born for whom their technology of self-expression has not yet been invented.

The End